A Pair of Shootists

A Pair of Shootists

The Wild West Story of S. F. Cody
and Maud Lee

Jerry Kuntz

University of Oklahoma Press : Norman

Library of Congress Cataloging-in-Publication Data

Kuntz, Jerry.
    The wild west story of S. F. Cody and Maud Lee / Jerry Kuntz.
        p. cm.
    Includes bibliographical references and index.
    ISBN 978-0-8061-4149-7 (hardcover : alk. paper)
        1. Cody, S. F. (Samuel Franklin), 1861–1913.   2. Lee, Maud Maria, d. 1947.
    3. Shooters of firearms—United States—Biography.   4. Wild west shows—
    United States—History.   5. Wild west shows—Great Britain—History.
    6. Aeronautics—Great Britain—Biography.   7. Air pilots—Great Britain—
    Biography.   I. Title.
        TL540.C64K86   2010
        629.130092—dc22
        [B]

                                                                    2010005962

The paper in this book meets the guidelines for permanence and durability of
the Committee on Production Guidelines for Book Longevity of the Council on
Library Resources, Inc. ∞

Copyright © 2010 by the University of Oklahoma Press. Published by the
University of Oklahoma Press, Norman, Publishing Division of the University.
Manufactured in the U.S.A.

1 2 3 4 5 6 7 8 9 10

CONTENTS

ILLUSTRATIONS

ACKNOWLEDGMENTS

My interest in S. F. Cody began as an offshoot of a biographical research project on the life of Alfred W. Lawson. Lawson was a baseball player/promoter who suddenly took up an interest in aviation in 1908. He said he was inspired to change careers after witnessing an airship floating over London in October 1907. It did not take me long to discover that the airship was piloted by S. F. Cody, or that Cody was an extremely interesting character. Seeking further information on Cody, I went straight to the best source: Jean Roberts, who lives in one of Cody's last residences and has become the most knowledgeable expert on his career.

Early in our correspondence, Jean emphasized that much was unknown about Cody's years before he moved to England, particularly the details of his first marriage. At the time, I had just discovered the first digitized historical newspaper products available from Cengage, Proquest, Newsbank, and Heritage Microfilm; out of curiosity I searched them to find material on Cody and his first wife, Maud Lee. The story that revealed itself through those news articles was too compelling to set aside. To Jean I owe great thanks not only for the source material she has provided, but for the encouragement she offered in tracking down elusive leads.

By the time I tracked Cody and Maud to Harry Hill's Oklahoma Historical Wild West Exhibition, I was frustrated to find absolutely no material on that show—until I discovered that a chapter on that show was soon to be published in Stan Hoig's book *Cowtown Wichita*. Stan graciously sent me his material in advance of its publication, and we subsequently swapped some anecdotes about Oklahoma Harry Hill. Stan took an interest in my research, and to him I owe thanks for urging me to expand it to book length.

ACKNOWLEDGMENTS

Another regular correspondent has been Ralph Finch, editor of the glass target ball collectors' newsletter *On Target!* Ralph has an encyclopedic knowledge of early sharpshooters, and was able to answer many questions I had about Bogardus, Carver, and others. Ralph also deciphered what Maud was trying to tell reporters about the "recipe" for composite target balls she purchased, and what the likely ingredients were. Moreover, he identified the ball molds used by Cody.

Others who deserve thanks for the information they offered include Paul Fees, consulting expert for the Buffalo Bill Historical Center and the Annie Oakley Center; Juti Winchester, of the Buffalo Bill Historical Center; Brenda Arnett, of the Annie Oakley Center of the Garst Museum; Manola Madrid, of the Autry National Center; and Susan Labry Meyn of Cincinnati, expert on the 1896 Sicangu Sioux encampment.

Lyrics from *Annie Get Your Gun* are reprinted by permission: "There's No Business Like Show Business" and "They Say It's Wonderful," both by Irving Berlin, © Copyright 1946 by Irving Berlin, © Copyright renewed. International copyright Secured. All rights reserved.

A Pair of Shootists

CHAPTER 1

THE ADVENT OF
THE SHARPSHOOTERS

1870s–1882

Samuel F. Cody was twenty-one and Maud Lee was sixteen when they first met in the early months of 1888. They were brought together by a shared ambition to attain Wild West show stardom, inspired by the fame of the main originator of that format, William F. "Buffalo Bill" Cody and by his leading female star, Annie Oakley. In the last fifteen years of the nineteenth century, the Wild West show brought worldwide celebrity to Buffalo Bill and Annie Oakley—fame then unprecedented for entertainers. S. F. Cody and Maud Lee fell short in their hopes to reach the level of public recognition for their riding and shooting skills that Buffalo Bill and Annie Oakley attained.

After setting aside the entertainment business, S. F. Cody refocused his ambitions and gained abbreviated success as a pioneer aviator. This fame soon faded as well, however, eclipsed by the pace of change in the new century. His accomplishments as a flyer were isolated successes that did not lead to a lasting business or to notable technological innovations.

History was even less kind to Maud Lee: her name appeared in a few news articles over the course of a dozen years, then she disappeared into oblivion. Yet the story of their few years together and the divergent paths their lives took after their separation is a tale full of exuberance and heartbreak—and a compelling counterpoint to the romanticized image of Annie Oakley and her life with husband Frank Butler.

Several events in the professional lives of S. F. Cody and Maud Lee can be understood only in the context of how Wild West show shooting acts came into being. The first "fancy shooters" to find a home in the

Wild West show format had performed in other settings in the years prior to Buffalo Bill's first arena show in 1883. These venues—made possible by innovations in quick-loading firearms—included shooting club and open shooting competitions, frontier-themed stage melodramas, prevaudeville variety-hall fare, and circuses. Many of the key figures from these traditions made the transition to Wild West shows, and frequently crossed paths with one another—and with S. F. Cody and Maud Lee.

The evolution of two aspects of Wild West sharpshooting—the use of glass target balls and the pedigree of the trick called the "mirror shot"— serve as means for introducing brief character sketches of these key figures. Target balls and the mirror shot loom large in the story of the two lovers, Cody and Lee, and so do several key sharpshooting personalities. Many readers will recognize the names of Annie Oakley and Frank Butler, Adam Bogardus, and Doc Carver. Frank Frayne was famous in his lifetime, but is now an obscure footnote; and the career of Jennie Franklin, who appeared on billings under various married and stage names, was never fully known to anyone in her lifetime or since, until now.

Adam Bogardus led a volunteer company in the Union Army for six months during the Civil War in 1863, but on his return home he found himself drifting away from civic interest in his adopted town of Elkhart, Illinois. The more time he spent hunting in the field, the less inclined he was to toil for a living in the sawdust of his carpentry shop. After facing the grisly tableaux of battle, Bogardus came back from the war and found that hunting game, his passion since he was a boy, was the one activity that brought focus to his life.

He excelled in handling a shotgun—as he had done since his boyhood days outside of Albany, New York. A few years after returning to Elkhart after the war, he forsook carpentry and turned to a different, uniquely American vocation: market hunting. He killed birds—grouse, snipe, duck, turkey, quail, woodcock—not for his own table, but for shipment to the restaurants of the booming cities of Chicago and St. Louis. He shot fowl in vast quantities, unburdened by nonexistent game limit laws. It was a lucrative and manly livelihood, and his stature in the community rose as he transformed himself from a mediocre carpenter to an expert gunman.

Inevitably, other shooters issued challenges to test their skills against Bogardus. He was tied to Elkhart by family and by its proximity to transportation routes to urban markets. Nonmigratory fowl were getting scarcer, however, and competition from other market hunters was getting more aggressive. Bogardus was not keen on accepting the first challenge he received: he did not like to best strangers, and shooting at pigeons released from traps did not appeal to him as a test of skill. In 1868 he accepted his first challenge to shoot against another man with a prize of $200 in the balance. He won by a score of forty-six birds to forty. In just a few minutes, he earned as much money as he usually did for selling about 1,300 birds to the restaurant trade. Bogardus wasted little more time on market hunting and instead started sending out shooting challenges of his own.

It did not take long for Bogardus to realize that in man-to-man competition, a little gamesmanship went a long way. He started to refer to himself as Captain Bogardus, a rank he earned in the war and an honorific that immediately commanded respect. He purchased expensively tailored attire to shoot in, and made a point of appearing immaculately groomed. He arrived at shoots with several expensive guns, stored in handcrafted cases. He often suggested matches where decorative badges were at stake along with the prize money, and posed for public photographs wearing those gaudy medals. Intimidated by Bogardus's bearing, some opponents might have felt they had lost the match before a single shot was fired.

As the 1870s progressed and the captain's collection of medals grew, he began to face two problems. First, he was running short of competition—he had beaten all comers, including state champions and the champion wing shot of Canada. Since the posted prizes of his matches were supplemented by numerous side bets among the crowd, the sport needed a semblance of competition. As a result, the shooting meets grew from a single skill test to multiple events, such as shooting two birds at a time or shooting against a clock or shooting from a moving wagon. To attract crowds, Bogardus even invented a special fireworks cartridge that could be fired from a rifle.

The second problem was that the size of the meets combined with the number of events required literally thousands of live, wild, target pigeons—passenger pigeons. There were still large flocks of passenger pigeons in the 1870s, but not as many as earlier in the nation's history, when passenger pigeons had been the most abundant bird on the planet. For trap shoots, pigeons were caught on the ground in nets after ingesting alcohol-spiked grain or after being drawn to a blinded live decoy tied to a little perch (a "stool pigeon"). The pigeons, once despised by nearly all white settlers as flying vermin, now had a defender in a recently formed national organization called the American Society for the Prevention of Cruelty to Animals. Bogardus and his cohorts were somewhat stunned to find that their pigeon shoots were not welcome in many eastern cities. This was irritating, because those cities were home to the high-stakes gamblers who thrived on such competitions.

Bogardus approached these problems as a challenge to his ingenuity. An alternative to live targets was needed in order for shooting competitions to draw crowds. Years earlier, glass ball targets had been developed in Britain, along with a special sling to hurl them, but they were heavy and could not be thrown more than thirty feet or so.[1] Bogardus commissioned the molding of lighter glass balls, patented a faceted design, and designed a new sling mechanism to throw them. He then staged shooting exhibitions at major urban venues to promote these innovations.

The result was a revolution in trap shooting. The easily shattered glass targets allowed breech-loading shotguns and light shot to be used, greatly increasing reloading speed and reducing the bruising effects of recoil. In March 1877, at Madison Square Garden in New York City, Bogardus shot one thousand balls in one hundred minutes. The glass targets breathed new life into wing shooting competitions and exhibitions, as well as attracting the attention of expert rifle and pistol shots, and a new form of popular attraction was presented to the American public: fancy rifle shooting.

Adam Bogardus's reign as the most famous shooter in America was short-lived. He received a challenge from California, from an upstart who claimed he could better all of Bogardus's feats and, moreover, could do so with a rifle rather than a shotgun. As he traveled eastward, William F. "Doc" Carver gave demonstrations of his astounding skill. Carver threw down the gauntlet while Bogardus was touring Europe, but by the time

Bogardus arrived back in the United States, Carver had left on his own three-year European tour. The match between the two champions would not take place until well into the 1880s.

Bogardus's refinement of the glass target ball was a major factor in the rise of celebrity sharpshooters. Even before that development, however, public performances by marksmen were seen on American stages and in tent shows. Because wing shots and moving shots were difficult to stage in theaters, indoor shows relied more on set trick shots. Onstage, they often took place within frontier-themed melodramas, such as those that William F. "Buffalo Bill" Cody starred in during the 1870s, prior to developing his Wild West show. Some shooting acts were presented without any dramatic context and were components of touring variety shows and circuses—a tradition that eventually found a home in vaudeville circuits and theaters.

The set trick shots demonstrated by these performers varied little in the twenty-five years between 1875 and 1900. The repertoire of every set trick shot act included some or most of the following feats:

- Snuffing out a candle with a bullet
- Shooting a pipe or the end of a lit cigar from the mouth or hands of an assistant
- Shooting and breaking a length of hanging rope (in dramatic contexts this often saved the life of a lynching victim)
- Bending forward and shooting between one's legs at a target behind
- Shooting apples, potatoes, or turnips off an assistant's head (sometimes the assistant was a trained dog)
- Leaning backwards over a chair and shooting with head and rifle upside down
- Shooting while held upside down, standing on one's head, or dangling upside down from a trapeze bar
- Shooting at a second gun rigged with an extension to its trigger, a hit on which would cause the second gun to fire—often at a target right next to or above the shooter.

- Shooting while holding the rifle or pistol sideways, so that the barrel could not be used to aim the shot
- Shooting at a target ball swinging from a string
- Covering the gun sight on the barrel of a pistol or rifle with a piece of cardboard, then shooting at a target
- Splitting a playing card in half through its edge

One trick shot in particular was so enthralling that it was usually saved for the climax of a performance: the mirror shot. The mirror, or backwards, shot requires the shooter to fire the rifle backward over the shoulder while aiming it by looking forward at a reflection of the gun sight in a mirror. The mirror shot has become an icon in popular culture, but few today realize it had a deadly history.

The man credited with the first public performance of the mirror shot was Frank Ivers Frayne, a showman who moved from trained animal acts into frontier-themed dramas. He first appeared onstage in western melodramas in 1874, but it was his 1875 production of the play *Si Slocum* that introduced his most spectacular shooting stunts. Frayne, a Kentuckian by birth, picked up his marksmanship skills in mining towns following the Civil War. In the climactic scene of *Si Slocum,* Frayne's character, Si Slocum, is forced by the villain to perform the mirror shot in order to secure the release of his wife, Ruth. The scene was made sensational by the fact that Frayne had to shoot an apple from the top of her head—an allusion to the legend of William Tell.[2]

Audiences were savvy to stagecraft that could fake shooting stunts via the use of blank cartridges and manipulated targets, but Frayne was known to use live ammunition, and people flocked to his production of *Si Slocum.* Often, Frayne would open a theater engagement in a town by giving a private demonstration to disbelieving members of the press. The play toured the United States for more than seven years, and played in Great Britain as well. The public was undoubtedly well aware that the role of the wife in *Si Slocum* was played by Frayne's real-life spouse, Clara Butler Frayne. To quell potential uproar over the perilous entertainment, Frayne issued public assurances that several precautions were taken during the mirror shot: Clara wore a hat that elevated the apple four to six inches above her head, and underneath the hat she wore a chain-mail skullcap with metal plates.

Jumping on the publicity that followed the opening of Frank Frayne's *Si Slocum* in 1875, other performers began to imitate Frayne's shooting tricks in their own acts. Among the first were the Austin Brothers, who in 1877, for a $125-a-week variety-theater salary, would take turns shooting at targets held by the other. The Austins also toured with Buffalo Bill's 1877 stage show *May Cody*. The mirror shot was a main feature of their program. The *Brooklyn Eagle* thought the act went too far:

> Does the pleasure of witnessing so nice an aim compensate for the possible spectacle of two corpses on the stage, and the sickening thought of brains spattered over the audience? . . . It is no consolation to know that there has never been an accident, for human calculations are not infallible, and as surely as this feat is attempted, it will be attempted once too often. The fatal performance may not take place tomorrow. At the same time it may. . . . Brooklyn audiences are not degenerate Romans whose holiday is incomplete without a gladiatorial butchery. There is no amusement in waiting to see two men drop dead.[3]

The inevitable fatal performance that the Brooklyn editor feared took place a little more than a year later, on April 6, 1878, in Pawtucket, Rhode Island. The shooter was a woman named Jennie Fowler, who used the stage name Jennie Franklin (to match the stage name of her husband). Mrs. Franklin, before performing as a trick shooter, first showed her mettle as a female boxer. She fought male boxers in the ring inside Harry Hill's saloon on East Houston Street in New York City. Harry Hill's was the most notorious establishment in nineteenth-century America, and the center of the sport of prizefighting. (Hill, a former wrestling champion who emigrated from England, should not be confused with Harry Hill of Oklahoma fame, who will be mentioned later.) Mrs. Franklin claimed she was an outcast member of a prominent Brooklyn family, but that fact remains unproven, as does her real maiden name.[4]

Because her regular assistant was ill, Jennie Franklin was aided by another member of the variety troupe, Louisa R. Maley, a trapeze artist who used the stage name of Mademoiselle Volante. During the performance that evening, Franklin displayed skill in hitting her target and other objects while firing directly at them, and she seemed confident and assured as she prepared for the final feat. Mlle. Volante stepped alertly to her station, smiling at the audience as she did so, and placed the apple on top of her head, where it rested in her luxuriant hair. Jennie Franklin also took her station near the footlights, in front of the mirror, and aimed over her shoulder through the looking glass at her ill-fated accomplice. The audience sat in silence, watching the performers, when suddenly as the trigger was pulled, a shriek resounded through the hall, and Volante fell forward on the stage. At first it was supposed that Mlle. Volante had merely been wounded, but within an hour the news spread that the woman had been killed, the fatal bullet having entered her forehead at the hairline and exited out the back of her head. The surgeons who were called immediately probed the wound, causing it to bleed profusely.

A month later, Jennie Franklin was back performing her act. The *New York Sun* covered her return to the stage and included a candid interview that sounds as compelling to a modern reader as it did 130 years ago:

> Mrs. Jennie Franklin, advertised as "the daring shot, and principal in the terrible catastrophe and innocent killing of a woman onstage in Pawtucket," stepped on the stage of Tony Pastor's theatre, at the matinee yesterday, gun in hand. She is five and a half feet in stature, and lithe and straight as an Indian. A blue velvet cap, knotted to her shoulders with ribbons, hung jauntily down her back, leaving her arms free. Blue short skirts, snow-white tights and blue leggings adorned the rest of her graceful figure. Before she shot Volante, a little more than three weeks ago, her cheeks were rosy and she weighed 157 pounds. Now she weighs 119 pounds and her cheeks are pale.
>
> Mrs. Franklin's twenty-nine-pound gun was held with the steadiness of a Creedmoor rifleman. Her husband fastened apples

on the face of a target, and she split them in two with bullets. She extinguished a lighted candle and broke a clay pipe. The range was the width of the stage. At last turning her back to the target, she pointed the gun over her shoulder, and taking aim over the reflection of the gun in a mirror, at the reflection of the apple—missed. She tried three times without success, and then bringing her gun to an order arms, with an angry pout, bowed and retired, the spectators heartily applauding.

"I could have hit the apple," she explained in the green-room, "if the audience had kept still. Just as I was raising the gun I could hear people say—'That's the shot she killed Volante in.'

"I had to miss then. I can hear what people say if it's only a whisper in the back of the room. Then some ladies in one of the boxes made some remarks about me. I want people to know one thing—that I never shoot below the mark. If I miss, the bullet always goes above it. I think Volante must have breathed so (drawing a long breath), after getting off the trapeze that night, and that threw her head up. I laughed when she fell, and sat on the floor with her hands on her knees; I thought she was making believe, as she had done before, to scare the audience. Then I said: 'My God!' I went over to her and I knew at once the bullet had come out again. I saw a few drops of blood under the hair just over her forehead. The wound in the forehead where the bullet went in didn't bleed. I wanted then to wash her head and let her go to sleep, but the doctor inserted a probe, which cut the principal vein that carries the blood all through the head. The blood came out in a stream, and it continued to bleed. They wouldn't believe me when I told them that the bullet had come out, but they had to afterwards. I attended her, and when her right side was paralyzed, she tried to talk to me, but she couldn't shape her lips to speak the words. She would take my arm in her left hand and draw me to her, and when I stood close beside her, leaning over her bed, she would put up her hand and raise her left eye so that she could see me.

"I am paying $5 a week for the board of a woman who used to stand for me [in the act]. She wants to do it again. I have to support

my child, and it [the shooting act] is all I can do. I can't save anything. You look at me here and you don't know me, but you would
know me well if I should tell you the name of my family in Brooklyn. That is a secret I have kept through all my trouble."[5]

Following that brief booking at Tony Pastor's theater in May 1878,
Jennie Franklin disappeared for six years. In the aftermath of the tragedy,
marksman Frank Butler wrote a letter to the editor of the *New York Clipper,* the leading theatrical paper of its day:

Since the shooting of one performer by another some few days
ago, I have frequently been asked, "What do you think of it?" and
as I have performed this same act frequently, in public and in private, I will endeavor to give my experience in rifle shooting. In the
first place, all performers make that shot with the use of a looking
glass because it is not so hard to make after you learn it as the
straight shot. This may seem strange, but it is so.

With the rifle held arm's length or in its natural position, a person holding it any length of time will be apt to shake more or less,
but by putting it over the shoulder and resting the butt, with the
glass resting also, the shooter can take time and with practice get
as good sight as in straight shooting.

Many newspapers lay the cause of the unfortunate accident referred to on the reflective power of the mirror. I do not think this
reason tenable. Others say that the woman, upon whose head the
apple rested, moved after the apple was placed in position. This
could make no difference, as the shooter, if she understands her
business, had to take aim at the apple, not at the place it had been.
With all due respect to Mrs. Franklin as a shooter, I do not think
she could have had good aim. [Or] she had a bad cartridge, and
right here allow me to say there are strange, very strange things
happening in shooting.

I have seen a cartridge, shot from a rifle at a distance of twenty
yards, drop six inches, and only go a half inch in a pine board,
while all the rest went through a three inch plank. This occurred
in the summer of 1876, while I was managing a shooting gallery

at Gloucester, N.J. Two gentlemen having seen me shoot an apple out of the hand of a friend, one volunteered to hold his hat while the other shot through it. At the first shot the man dropped the hat, and we found he was shot in the hand. The man who fired the rifle was an excellent shot.

One of the strangest things that ever happened under my observation occurred last summer at Gloucester. I had shot hundreds of apples and potatoes from my dog's head without a miss. The dog I placed on a table facing me, with the potato on his head. This time I shot and split the potato to pieces, and then found that my dog was shot in the back of the neck. Had he been shot in the face or head, or in any exposed place facing me, I should have blamed myself. As it was, several gentlemen examined the potato and found several small pieces of shell and stone imbedded in it, and the theory they put forth was that the bullet struck a stone or shell which changed its course, and, instead of going straight through, turned down and went into the dog's neck, came out through the shoulder blade, and then went through a one inch board.

Some few months ago Mrs. Frank Frayne shot her husband's fingers off, yet she had been shooting apples out of his fingers for ten years without an accident, and she says her aim was as good that time as it ever was. Had she been shooting an article off his head, he would have been the victim. Then there were two shooters, one of whom held a block of wood in his mouth while the other shot into it; but one day the holder of the block got shot in the chin, and now he doesn't hold blocks any more.

In my opinion this kind of shooting can be done without endangering the life of anyone, although the more danger there is the higher salary the average manager will pay. In conclusion, let me advise all shooters to exercise the greatest care in the use of firearms; never point a gun at another, even though you could swear it was not loaded. Accidents will happen. F.B.[6]

Phoebe Ann Mosey could often be found stalking the western Ohio woodlands, her dogs padding alongside. She was a talented wing shot, but credited the dogs with much of her success as a market hunter. In 1875, at age fifteen, she was already a calculating businesswoman. The income from the game birds she harvested had already paid off the $200 mortgage on her mother's home. Phoebe (who went by her middle name, Annie) had begun working towards that goal only a few months earlier, on her return to her family's household from the county poor farm.

When Annie was ten, necessity had forced her mother to send her to the poorhouse as a ward. Those years of banishment included a long stretch of servitude to a local farmer and his wife, who treated her cruelly. Annie had learned sewing and embroidery at the poor farm, but hunting earned quicker profits than needlework. She became adept with the rifle in just a few weeks.

Most people in Greenville, Ohio, knew about the hardships Annie had endured, so their admiration for her determination outweighed the view that her vocation was inappropriate for a woman. Annie herself allowed no opening for that criticism: she always dressed in long skirts, never behaved improperly, and invariably presented a cheerful demeanor. Mr. Katzenburger at the local general store bought her bags of quail, grouse, and woodcocks and made a fair profit from them.

How long Annie toiled as a market hunter is now a subject of contention. In the late 1870s she, like Adam Bogardus, must have faced the pressures of competitors and decreasing game, not to mention the lure of easier money in competitive shooting. In later years many contradictory claims surfaced about Annie's activities between 1875 and 1880. The intentional obscuration of her history could have been the result of any of several motives: a desire to subtract years from her age, a whitewashing of a romantic liaison that was awkward in its details, or an attempt to disguise the social impropriety of a single woman, acting independently and ambitiously, taking up show business as a career.[7]

It is known that by the middle of 1880, Annie had moved from market hunting to exhibition shooting and begun to appear on the bill of variety-hall theaters in Cincinnati, Ohio. She took the stage name "Annie Oakley," most likely from a neighborhood in Cincinnati called Oakley. During these shows, she performed as a solo act. In later years Annie may have wanted to hide this history, because a single woman working in variety halls would not have been considered respectable. Above all else, she valued her good reputation.

By her preferred account, Annie met her future husband, Frank Butler, at an arranged shooting match. In various retellings of her story, however, the year and the location of that match varied. Frank had begun touring in variety halls with a veteran marksman partner named Samuel Hyde Baughman a few months before Annie first appeared on stage. In February 1881, both Baughman and Butler and Annie Oakley appeared on the same bill at a theater in Cincinnati. This is the first documented meeting of Annie and Frank.

Baughman and Butler joined the Sells Brothers Circus in April 1881. Less than a month into the touring season, the shooting risks that Frank Butler had forecast caught up with him. One particular part of the act involved Frank shooting a potato or turnip off the top of Baughman's head. On the night of May 23, while playing in Milwaukee, Butler appeared to the spectators to be very agitated. He shot at the turnip on Baughman's head and missed. He shot four more times in rapid succession, missing each time. Finally, on Butler's sixth shot, Baughman pitched forward and fell on his face. The small-caliber bullet had penetrated above his right eyebrow to the bone, but then glanced upward, leaving no more than a deep scalp wound. Baughman survived, but their partnership did not. On regaining consciousness, Baughman's first words were a request that his mother not be told about the accident. He was confined to bed for two months. Frank continued with the Sells Brothers Circus with a new partner, John Graham.

Following the circus season, Graham and Butler returned to the variety-hall circuit. They met Annie again in Cincinnati in January 1882. Annie Oakley went on tour with Graham and Butler, appearing on the same bill (as a separate act) as they went through Indiana, Ohio, Pennsylvania, and Washington, D.C.

A year after Baughman's accident, in May 1882, Frank's new partner John Graham fell ill and could not dress for the curtain. Frank tried to perform solo, but shot badly and had the audience heckling. Since Annie was traveling with Graham and Butler, she took Graham's place. Frank never sought another partner. Once he observed what a crowd pleaser

Annie had become, he allowed her to become the star of their act, and relegated his own role to assistant. They married in June 1882, and spent the next year once again touring variety halls and enlisting with the Sells Brothers Circus.

In the five years that Frank Frayne's wife, Clara Butler, performed in *Si Slocum,* she suffered no injury during the hundreds of times she held the apple for the mirror shot. However, during another shooting set shot in which Frayne fired at a target held from her hand, she lost a finger. Despite that mishap, *Si Slocum* made a small fortune for the Fraynes, and in the spring of 1880 they had an elegant new house built in Chicago. Clara, however, hardly had time to enjoy their new home; she died from asthma just two weeks after they moved in.[8]

Frayne continued touring with *Si Slocum,* with Clara's part in the cast being taken by another actress for a short while, then later in 1881 by twenty-five-year-old actress Annie Von Behren. Within a year, Von Behren and Frayne were engaged to be married. *Si Slocum* opened in Cincinnati, Ohio, in November 1882. The Thanksgiving matinee performance on November 30 was well attended by women and children. In the climactic scene, the arrangements for the mirror shot were made, the rifle cracked a report, and the character of Ruth Slocum fell to the floor without uttering any audible sound. Frayne looked around in a dazed way, looked down at his shirt, which had been singed, then rushed towards the fallen Annie Von Behren, crying, "My God, my God, what have I done!" and fainted beside her. The curtain was immediately dropped. The audience was still unaware that this was a departure from the script.

Frayne had been using a breech-loading Stevens rifle with a barrel that folded down from the stock for loading the cartridge. When the barrel was straightened back up, a snap catch secured it in position. On this occasion, the catch gave way as the shot was fired, which caused the barrel to drop below its aim; at the same time, the flash from the cartridge blew out of the breech onto Frayne's shirt. The ball hit Von Behren in the forehead above the left eye. The entrance wound was a large hole, and as it bled, pieces of her brain oozed out. She never regained consciousness and died within fifteen minutes.

Frayne, once he was revived from his swoon, was arrested. After a preliminary hearing, he was released on $3,000 bail. He was promptly brought to trial on the charge of manslaughter, supported by a statute that outlawed the pointing of a loaded gun at another person. Frayne's defense counsel countered that the rifle was not pointed at Miss Von Behren, but was instead pointed at an apple several inches above her head. The judge agreed, and furthermore noted that Frayne, who broke down in court, had suffered enough.[9]

Frank Frayne immediately announced he had sold his shooting apparatus and was retiring from the stage. The next year, however, he did appear in another production, without guns. Instead, he added thrills to the production with animals, including a wild dog, a bear, and a den of lions. By 1886, he was back on stage, doing shooting stunts in the *New Si Slocum,* although he did not shoot at targets held by people. No more mishaps appeared to plague Frayne's career; he died in 1891 of heart disease at the relatively young age of fifty-two.

These disparate episodes illustrate that sharpshooting was a firmly established performance act by the early 1880s. By that time several marksmen had already gained celebrity, even notoriety. Risky stunts such as the mirror shot threatened to give set shooting the reputation of a blood sport, while trap shooting was invigorated by the use of artificial targets. The one element lacking from shooting performances was a sense of grand showmanship. That void would soon be filled through the inspiration of William F. "Buffalo Bill" Cody.

CHAPTER 2

THE BIRTH OF
THE WILD WEST SHOW

1883-1887

The first showdown between the two champion shooters, Captain Adam Bogardus and W. F. "Doc" Carver, did not take place until February 1883. Both men had returned from their successful European tours long before that date, but Bogardus felt he had little to gain from risking his recognized crown as champion shooter. Carver's challenges grew more strident, however, until they reached a point where Bogardus's pride could not accept being labeled an "artful dodger" by Carver any longer.

Their first match was held in Louisville, Kentucky, on February 22, 1883. They agreed to use shotguns in deference to Bogardus, with live pigeons as targets. Bogardus's intimidating presence—the elegant clothes, the custom guns and cases—had no effect on Carver, who was 6' 4", weighed 265 pounds, and wore an outlandish western outfit of a fringed buckskin coat, a huge wide-brimmed cowboy hat tilted high above his brow, and leather leggings decorated with glass bead designs. Bogardus's title of "Captain" was offset by Carver's "Doc," though Carver's credentials as a dentist were lapsed. By some accounts Carver's early adulthood was spent in dentistry, though in later years he implied his family gave him the nickname of "Doc."[1]

Carver won the contest, eighty-three birds to eighty-two.

Bogardus requested a rematch, and Carver eagerly agreed. They met next in Chicago, and this time shot one round of one hundred live birds and a second round of clay targets. Carver prevailed in both rounds.

The clay targets were a new innovation, and even Bogardus had to acknowledge that they flew better than his own patented glass-ball system. The glass-ball targets were still used for many years afterwards, however, because exhibitions in enclosed buildings and around large crowds required the use of shells with light loads of gunpowder, and glass balls would break when shot by light loads whereas other targets would not.

Thousands of people attended these first two matches, and the results were printed in newspapers across the country. Bogardus and Carver agreed to meet again in St. Louis. On their arrival there, they received a written offer from George Ligowsky, the maker of the clay targets they had used in Chicago, to compete in a whole series of twenty-six matches, with the winner to take $300 per match. These contests took place in March and April 1883 across many eastern cities, with very consistent results: Carver won nineteen times, Bogardus three times, and three times they tied.[2]

The popularity of the Bogardus-Carver matches served to solidify the plans of one of Carver's old prairie hunting acquaintances, William F. "Buffalo Bill" Cody, to promote a traveling exhibition of scenes and skills from the American frontier. Buffalo Bill had already made a name for himself in the late 1870s with stage melodramas built around his larger-than-life character, an image derived from fanciful Ned Buntline dime novels. Cody had found that indoor stages limited the action he wanted to present. His idea was to design a show for an outdoor arena that would feature authentic white settlers, American Indians, cowboys, scouts, soldiers, wild horses, cattle, elk, and bison in skits demonstrating fighting, roping, horsemanship, hunting—and, of course, shooting.[3]

Doc Carver bought into Cody's grand plan and swore himself as Cody's partner, even as Carver was still touring with Bogardus in the spring of 1883. Cody and Carver bandied different names for the show back and forth, settling on "Buffalo Bill and Doc Carver's Wild West, Rocky Mountain, and Prairie Exhibition." Not only did Carver invest in the show, he insisted on a starring role. The profitable success of the matches with Bogardus (even if the outcome was lopsided) enticed Cody and Carver to invite Bogardus to join the cast. Bogardus agreed, but wanted to build a separate act around his three young sons, who were already skilled shooters. The "Wild West" show concept came at the perfect time, when the great circus showman P. T. Barnum no longer seemed an innovator and when there was a national passion for the "vanishing West." The only thing lacking in the planning of Buffalo Bill and Doc Carver's Wild West Exhibition was the recruitment of a manager who knew something about the entertainment business.

The lack of show business acumen nearly doomed the tour in its first year. Typically, the show would sell out the initial dates in a town, but as it exhausted the audience of newspaper boys, shoeshine boys, and other dime-novel readers, it found few middle-class ticket buyers. To many city dwellers, the all-male cast appeared more threatening and barbaric than heroic. It did not help matters that Cody and nearly all the other noted cowboys on the tour used their time together as an excuse to drink copious amounts of alcohol. One entire car of their special show train was rumored to be a liquor warehouse.

Doc Carver may or may not have been one of the drinkers, but he proved to be a lackluster and temperamental performer. On several occasions he shot badly and took out his frustration on his mounts, abusing them in plain view of the audience. Cody had enough sense to realize that the partnership with Carver was not working out. He and Carver seemed to bring out the worst qualities in each other. At the end of the 1883 season, Buffalo Bill unilaterally ended the relationship with Doc Carver and implored a successful theater manager, Nate Salsbury, to be the show's business manager for the next season. Salsbury signed on and promptly put an end to the open binge drinking.

Salsbury immediately grasped that the company needed more elements of traditional showmanship in order to appeal to a broader audience. Cody had an aversion to the word "show"—he thought "exhibition" conveyed more authenticity—the eventual compromise was simply Buffalo Bill's Wild West, with neither "show" nor "exhibition" in the name. But Salsbury recognized the Wild West performance for what it was: a scripted entertainment. He added a band to heighten the dramatic impact of various skits; and he developed the show's trademark climactic scene, the raid on the settler's cabin. Through 1884, Salsbury's efforts stopped the hemorrhaging losses the show had incurred, but it was not yet profitable. Some element was still missing that Salsbury could not quite put his finger on.

The grind of travel on the Sells Brothers Circus schedule wore on both Frank Butler and Annie Oakley. They lasted through the 1884 season, but after the final engagement in New Orleans, they solicited Buffalo Bill Cody to join his Wild West show. Although still not solvent, the Wild West show was gaining a reputation in the entertainment business as a top-notch operation—the Butlers could reach no higher than to be a part of Cody's show. Cody declined their initial overture, however; although Doc Carver was gone, between his own shooting and that of Captain Adam Bogardus, he felt the show had an ample roster of sharpshooters. Frank and Annie resigned themselves to a return to variety-theater performances.

As Buffalo Bill's Wild West prepared for its first performances of 1885, it found itself reeling from a major mishap: steamers carrying much of the equipment of the show—most of its animals and Adam Bogardus's rifles—sank in the Mississippi. Bogardus quit his contract in a dejected huff. Cody suddenly was in the market for a sharpshooter, so Annie and Frank reappeared to audition. Nate Salsbury was present at Annie's tryout, and immediately recognized not only her skill but her appeal. She was hired on the spot, and Frank continued as her manager and assistant. Annie was an immediate hit with the Wild West audiences: more than 140,000 people in forty different cities saw her perform in 1885.

In 1886, Buffalo Bill's Wild West played the entire summer in New York, encamped at Staten Island. In that season they drew half a million people, and Annie's name was billed second only to Buffalo Bill himself. Even so, Cody felt compelled to add another female shooter to the show for their trip to England in 1887—a fifteen-year-old prodigy named Lillian Smith. Most likely, Salsbury and Cody wanted to prevent Lillian from signing with another Wild West show or circus. Their success had now bred many competitors, and Lillian had already made a name for herself in nationally publicized challenge matches. Lillian was profane and brash, anointing herself as a "Champion" shooter. She brazenly announced that she would be the show's premier shooter. She also partied with the show's cast of cowboys, while Annie and Frank remained aloof. To put it mildly, the two women sharpshooters did not get along.

Annie Oakley tolerated Lillian for one season only—and finally bested her in a long-delayed head-to-head match. After the tour of England completed its run in the fall of 1887, Annie and Frank quit Buffalo Bill's show. The next summer she toured with Pawnee Bill's Historic Wild West show and also accepted challenge matches at state fairs. Annie coexisted without incident with sharpshooter May Lillie, Pawnee Bill's young wife. Towards the end of 1888, Annie agreed to star in a new western-themed melodrama stage production that Frank decided to promote, *Deadwood Dick.* The show called for a large cast of cowboy heroes and bad men, which, since the Wild West shows were on their seasonal hiatus, were in short supply on the East Coast at the onset of the winter season.[4]

One handsome young cowboy was more than glad to join the production, however, considering the predicament he had just escaped. His adopted stage name was Samuel F. Cody. To explain how S. F. Cody was in a position to join the cast of *Deadwood Dick,* we need to pick up the trail of Buffalo Bill's erstwhile partner, Doc Carver.

Doc Carver's nickname was the "Evil Spirit of the Plains," which, according to his publicity flyers, was a translation of a title bestowed on him by Spotted Tail of the Brulé Sioux, who was supposedly awed by Carver's trophy kills of a white buffalo and a silver elk.[5] Before he issued his challenges to Adam Bogardus in the late 1870s, Carver's name is found in the lists of San Francisco gun meets as a good competitor—but often not the best.[6] There is no denying that by the early 1880s, Carver was an exceptional shooter, but some argue his rise to that status was based on countless hours of gun club range practice—not to skills honed on the wild frontier, where buffalo and elk did not fly on the wing.

Following his acrimonious breakup with Buffalo Bill at the end of the 1883 season, Carver spent the summers of 1884 and 1885 traveling with his own small-time Wild West show. He had frequent legal spats with Cody over the name "Wild West," which both men claimed to have originated. Carver was an even worse businessman than Cody, however, so by July 1885, his own show had folded. He spent the rest of 1885 and 1886 performing exhibitions and ducking challenges from the prepubescent wonder Lillian Smith—who hounded Carver the way he had hounded Bogardus.[7]

In 1887, Doc Carver teamed up with impresario Adam Forepaugh to headline the Wild West segment of the Adam Forepaugh Circus. Well aware of the new threat to his audience share from Wild West shows like Buffalo Bill's and Pawnee Bill's, Forepaugh was the first circus owner to incorporate a full-blown Wild West show into his tent show, and he hired Doc Carver to manage that part of the combination. Carver performed his sharpshooting exhibitions and also displayed some of the set trick shots he had mastered a decade earlier—including the mirror shot. Carver also embraced the role of the gallant General Custer in the circus's climactic Wild West skit, "Custer's Last Rally." Carver's preening nature well suited him for the starring role in the miscalculating general's death scene.[8]

In May 1884, newspaper reports surfaced detailing the exploits of a "Jersey Amazon." Jennie Franklin, driven by shame from the stage in 1878, had since remarried to a poor woodchopper named Jackson Moore, and they lived with Jennie's daughter Josie, now twelve, in a dugout log cabin in the wilds of the New Jersey Pine Barrens, near Dennisville, Cape May County. Jennie Moore admitted to reporters she had once performed as Jennie Franklin, but those writers had no memory of the Volante killing. In the spring of 1884, Jennie gave a demonstration of her shooting skills at the town hall in Dennisville, New Jersey, astonishing her neighbors. During these shows, she performed the mirror shot—with doughty Mr. Moore placing the apple on his head.[9]

Over the next year, Jennie likely became envious of the fame and good fortune that were now being showered on lady shooters like Annie Oakley—the new star of Buffalo Bill's Wild West—and Lillian Smith. With her daughter Josie now a teenager, the bucolic retreat offered by Jackson Moore lost its charm for Jennie. She left Mr. Moore and the Pine Barrens and remarried again, this time to a circus showman named William

B. Kennedy. Together, they forged Jennie a new identity to obscure the history of Jennie Fowler/Franklin/Moore, and to avoid any reminders of the killing of Volante. She became "Mexis," the female vaquero and Mexican equestrienne.[10]

Her pairing with Kennedy was brief; by 1888, they had separated, and Mexis joined the cast of the Adam Forepaugh Circus.[11] Forepaugh's 1888 Wild West cast thus had two seasoned practitioners of the mirror shot, Doc Carver and Mexis. On tour they would meet a young cowboy named S. F. Cowdery.

CHAPTER 3

S. F. Cody and Maud

1888

Adam Forepaugh, for his part, was delighted with the thrilling stunts the cowboys brought to his show in its 1887 season, but he could not resist making his circus bigger and better. To expand his Wild West show for 1888, Forepaugh committed $60,000 towards the acquisition of more Indians, cowboys, cowgirls, horses, buffalo, elk, and other entertainment. Over the winter, he assigned his agents to scour the land for prime specimens of a vanishing and half-imaginary way of life.

Adam Forepaugh was not a showman by training; he had been a cattle drover and horse trader who had acquired the assets of a small circus from a debtor, then built it into an entertainment empire second only to P. T. Barnum's.[1] Forepaugh and Barnum had their own history of cutthroat skirmishes, highlighted by 1884's "white elephant" war. Barnum had found an albino elephant and announced he was importing it to America. Forepaugh unveiled to the public his own white elephant, named the Light of Asia, before Barnum's elephant arrived. Barnum immediately labeled the Light of Asia a fraud. Forepaugh countered the charge by parading out expert testimony from university zoologists attesting that the Light of Asia was a genuine wonder. After a year, a former trainer finally revealed that the Light of Asia was an ordinary gray elephant that had been whitewashed with a plaster of paris solution.

After that episode, which hurt the reputations of both men, Forepaugh and Barnum called an uneasy truce. For the year 1888, they agreed to route their shows to different parts of the country. Forepaugh would tour the Northeast: eastern Pennsylvania, New Jersey, New York City,

Connecticut, Massachusetts, Rhode Island, Maine, New Hampshire, Vermont, upstate New York, and western Pennsylvania.

Adam Forepaugh was a ruddy, heavy man of Pennsylvania Dutch parentage. He had massive shoulders, thick legs, and muscular, sinewy arms. His hands were those of a workman, and Forepaugh more than once bunched them up and exercised them on employees who challenged his authority. Although he trusted his underlings to come up with publicity stunts, he himself took charge of what he called "faking the fakements."

Before Forepaugh took over as the "main guy" of his circus, one of his enterprises had been to supply horses to the streetcar companies of Philadelphia. He bought their old, tired nags, revived their spirits by giving them the free run of an island on the Schuylkill River, then resold them to a different company. The streetcar companies were amazed that the new stock seemed to have an innate sense for street traffic.

The winter headquarters of the Adam Forepaugh Circus were in a huge, high-ceilinged warehouse in Philadelphia. For a month and a half prior to the season opening in mid-April, the circus performers and Wild West cast developed their acts in the building. Newcomers, both human and animal, underwent training. Veterans honed their skills; prepared their costumes, scenery, and equipment; and repaired worn canvas and ropes. Inside the warehouse was a circus ring, in the center of which stood a huge wooden mast. From this mast a wooden armature was set on a rotating collar about twenty-five feet off the ground, allowing it to sweep the full circumference of the ring. Pulleys and ropes attached to it harnessed the trick riders as they ran through their drills of standing, leaping, and somersaulting.[2]

First, the trainees would go through their paces on the ground; only after mastering their moves on solid ground would they be allowed to try them from horseback. The horses used by the trick riders were white—not to impress, but to provide a contrasting visual reference for the leaping and twirling acrobats who rode them. The circus ring itself—its size and shape—was designed by trick riders to take advantage of centrifugal force to help them keep their balance.

Adam Forepaugh embraced equine display as the main theme of his show. The headline equestrian act was Adam Forepaugh, Jr., who also managed the other equestrian acts. For many seasons the climactic event of each show was his solo control of thirty bareback riderless horses from the back of one horse. He would direct the horses to walk up steep inclines, jump through papered hoops, and jump through rings of fire. The most famous of Forepaugh's horse acts was Eclipse, billed as the "trapeze-leaping pony," which had been trained to leap from one elevated platform hanging from chains to a second similarly suspended platform. During his seasons with Forepaugh's circus, Doc Carver never ceased to wonder at the appeal that the jumping horse had to the public, and filed away that impression for a future day. Another acrobatic horse of the circus was Blondin, the tightrope-walking horse.[3] Forepaugh also had another equine that he billed as the "biggest horse in the world." Thoroughbreds and jockeys re-created races around a quarter-mile track that ran the perimeter of the tent's two rings. Horse-drawn Roman chariots were an element of the set of "Sports of the Roman Amphitheater." To many audiences it must have appeared that the Adam Forepaugh Circus was one horse act after another.

In that spring of 1888, one trick rider trainee was a girl just turned sixteen named Maud Maria Lee. She hailed from the nearby town of Swedeland, near Norristown, Pennsylvania. Maud was the only child of English emigrants Joseph and Phoebe Lee. Joseph Lee was a foundry worker, originally from the industrial city of Manchester, England. He now worked in one of the many iron foundries near Norristown, where hazardous conditions were rewarded with meager pay. From a young age Maud fell in love with acrobatics and circuses. Maud was attractive, with a pleasing face and long brunette hair. She was about 5' 4" tall, and weighed about 120 pounds—nearly the same size as Annie Oakley. Her shoulders were broad and her limbs were firm from exercise. In street clothes she was described as "stout," a term which at that time conveyed health and vigor more than heaviness.[4]

Everyone in a circus had to perform multiple tasks to earn their wage, so aside from gymnastic schooling, Maud would have been taught other skills, such as working with the animals or donning skimpy Oriental outfits for the salacious "educational" skits. In the 1888 season, these skits included Cleopatra and her handmaidens trolling a golden barge to meet

Marc Antony; and Lalla Rookh leaving the court of her Mughal father to meet the mysterious prince to whom she is betrothed.

Female performers were also needed for the Wild West scenes. Cowgirls were made a necessary part of a Wild West show according to the precedent set by Buffalo Bill and Nate Salsbury. Genuine western cowgirls were as hard—or harder—to find during a Philadelphia winter as cowboys, so a trainee like Maud might have been needed to assist in practicing the western scenes.

Forepaugh's search for cowboys had reached far-off Helena, Montana, where a twenty-one-year-old horse trainer named Samuel Franklin Cowdery answered the call. Cowdery had left his Davenport, Iowa, home as a boy, after his ailing father had already deserted the family. Despite his relative youth, Cowdery was a veteran cowpoke by 1888; by his own account he was engaged in buffalo hunting, horse training, cattle herding, and prospecting since he was fourteen.

The Texas cattle ranches he later claimed to have worked on were in the eastern section of the Texas Panhandle, and had summer feeder ranches along the Powder River in Montana.[5] The longhorn cattle were driven north each spring, following the Western Trail through Dodge City and Kansas, Nebraska, South Dakota, and Wyoming. In later years, Cowdery told tales of his family's Texas homestead, Indian attacks, and prospecting for gold in the Klondike—tales that were implausible if not outright fabrications but echoed the self-mythologizing that nearly all Wild West showmen dabbled in to match the fantastic tales found in dime-novel fiction.

The only documented evidence of Cowdery's western exploits is a November 1882 item from a Deadwood, South Dakota, newspaper. It mentions that S. F. Cowdery left town on the Sidney-Deadwood Trail, one of the main trails heading south. Whether he had been herding cattle, training horses, or prospecting was not mentioned. He was, however, fifteen years old and traveling alone. Cowdery emerged from the 1880s in possession of undeniable proofs of his frontier education: he was expert in riding, shooting, and roping.

Such skills were the staple of Forepaugh's Wild West show. Aside from the main circus, the Wild West show alone had a cast of two hundred, including cowboys, cowgirls, sharpshooters, trappers, scouts, plainsmen, vaqueros, and cavalrymen. Forepaugh's "genuine blanket Indians" were a conglomerate of Sioux, Comanche, Kiowa, and Pawnee tribe members. The presence of American Indians in the cast was publicized using the then-current vocabulary that now sounds stereotypical and offensive: "bucks, squaws, papooses, chiefs, and medicine men."

In the late 1880s and 1890s, participation in Wild West shows by American Indians from the Plains tribes was discouraged by the federal government, ostensibly because of reports of their mistreatment in Buffalo Bill's show. A more likely motive was that federal Indian Bureau policies sought to force white culture on the reservation Indians—to make them dress, speak, work, and farm like white settlers. The U.S. government did not want the Wild West shows to perpetuate what it considered a barbaric and obsolete way of life that inevitably led to clashes. Therefore only a limited number of American Indians from the Great Plains were allowed to go on show tours—a problem addressed by Forepaugh in April 1888, when he engaged seven members of the Seneca tribe from western New York; their knowledge of Plains Indian culture was likely deficient, but in Forepaugh's eyes, they looked the part.

The circus performance script called for the cowboys and cowgirls to do a set of "Sports and Pastimes," including bronco busting; steer roping; horse roping; steer riding; and trick shooting with rifle, shotgun, and pistol. While the whites used firearms, the Indians performed demonstrations of bow shooting. The Sioux staged a "Sun Dance"—a tame re-creation of an actual ritual involving painful self-mutilation. Other scenes depicted were solders undergoing drills in a fort, the rescue of a white captive about to be burned at the stake, and the capture of a horse thief by lasso and his lynching.

One of the two extensive skits was the "Atrocious Mountain Meadow Massacre," a re-creation of the 1850s slaughter of white settlers by Southern Paiutes and Mormons disguised as Paiutes— presented by Forepaugh with all possible bias against the Mormons. The other climactic Wild West skit was "Custer's Last Rally," Doc Carver's showcase. A Little Bighorn reenactment had once been an element of Buffalo Bill's Wild West, but was later dropped. Several other scenic sets in the Forepaugh

Wild West show were direct copies of elements of Buffalo Bill's Wild West: the attack on the Deadwood Coach, Pony Express riding, and a Virginia reel danced by horses ridden by cowboys and cowgirls.

Upon his entry into show business, Samuel Franklin Cowdery changed his name to Samuel Franklin Cody. This was an obvious ploy to capitalize on his physical resemblance to a younger Buffalo Bill Cody, especially his long brown locks that fell over his shoulders. It is not known when and how the name change decision was made, but since Forepaugh was demonstrably trying to imitate Buffalo Bill's Wild West, and Forepaugh and his lieutenants had a history of concocting frauds, it seems likely that the suggestion was made to Cowdery by his circus bosses. Throughout his show-business career, Samuel F. Cody would alternately embrace and deny stories of his being a direct blood relative of Buffalo Bill, but he never went back to the name Cowdery.

While honing his act at Forepaugh's winter headquarters in March 1888, Samuel F. Cody met Maud Maria Lee. Maud was assigned to serve as Cody's assistant during practice of his trick pistol-shooting act. Her job was to hold glass balls—like those Bogardus had developed—in her fingers and to dangle them from her mouth as she was suspended in the air, while Cody shot at them.[6] The act required exceptional trust between the sixteen-year-old girl and the twenty-one-year-old cowboy. Soon after meeting, the two became—in the words of Maud's father—"kinder thick."

In her training to be a rider, Maud would have been instructed in the science of falling. Circus riders risked severe injury because their stunts depended not on their own skill alone, but also on the animal's steady pace. Any shying or hesitation by the horse would cause a loss of balance by a rider attempting any sort of move. During a performance, horses could be startled by any number of things: a crash of music from the band, the creaking of a pulley, a paper program falling into the ring, or noises from the other animals, to name a few. Veteran circus riders, if they start to fall, immediately grab their knees and tuck their head down, forming their body into a tight ball, with muscles tensed. If they are aware enough, they will try to hit the ground with their shoulders. It is a difficult skill to learn because one's instinct when falling is to throw out one's limbs to steady oneself.[7]

Maud did not travel with the show after it opened its 1888 season, either by choice or perhaps because she was considered too young and inexperienced. Forepaugh's six-month season for the year 1888 began with a week of performances in the circus's winter home, Philadelphia, Pennsylvania, starting April 23. Maud did go visit the circus when it came through Norristown on May 4, chaperoned by a friend of her father's who was given strict instructions to return from the circus with Maud in tow.[8]

The large cities on their route that year—Philadelphia, Boston, Pittsburgh—had weeklong runs, but all the other cities they visited had one-day engagements, with Sunday as the only day off from performing. There were two shows a day, in the afternoon at 1:00 P.M. and in the evening at 7:00 P.M.. Early each morning, the canvas men would erect the big tent and the seats while the performers rested in the cars of the circus train. At about 10:00 A.M., the performers and animals would parade from the railway to the fairgrounds, taking a route along the main street of the town. The circus parade was often as memorable as the performances themselves.

Adam Forepaugh liked to use the parade and the show's themed sketches to respond to accusations that circuses were frivolous, immoral, and corrupting. His solution was to use some sort of historical pretext to frame the acts. So, in addition to Cleopatra and Lalla Rookh, the circus included jousting medieval knights, mounted samurai, and chariots of the Roman hippodrome. A prize artifact that Forepaugh liked to show off during the circus parades was a dilapidated old wooden canoe, which he claimed had been used by George Washington during the French and Indian War.

The parade was a great teaser for the animal acts. Some of the draught horses used to pull carts to the tent were also used in the show, and the elephants were walked through the streets, led by Bolivar, the huge pachyderm that Forepaugh had purchased to rival P. T. Barnum's Jumbo. Jumbo was taller, but Bolivar weighed one thousand pounds more. Other famed Forepaugh elephants were Sullivan the boxing elephant and Picaninny the clown elephant (the racism of the era even seeped into the animal acts). In the show, elephants were also raced and made to dance, play music, and make a pyramid atop one another. To create a flashy display in the opening parade, Forepaugh had all the animal cages painted vermilion red—a practice that became instilled in circuses from that point forward.

The parade was also a direct source of income. Pickpockets worked the crowd. Some were unaffiliated camp followers, but Forepaugh had a reputation of working with some of them for a share of their proceeds. Frequently, circus parades were also followed by reports of articles being stolen from clotheslines in residential yards. Forepaugh's ticket booths were also noted for shortchanging their customers, and it was common knowledge that the sideshow games of chance were rigged. Although P. T. Barnum never actually did say "There's a sucker born every minute," that attribution was spread by Adam Forepaugh, who took the sentiment behind that phrase to heart more than Barnum himself did.

Crimes and accidents occurred during every circus tour, and Forepaugh's 1888 season was no exception. On June 13, in Springfield, Massachusetts, canvas man Harry Taylor, while working as an usher, punched an unruly drunk who would not move away from a reserved section. The man fell, broke his neck on the metal head of a tent stake, and died. Taylor, the son of the boss canvas man, fled the scene.[9] On June 21 in Webster, Massachusetts, three American Indians were knocked from the top of a parade cart. One was killed and the other two seriously injured.[10] On June 25 in Boston, Massachusetts, Oglala Sioux members of the show refused to occupy their assigned railcar after being told the same car had been used the year before by Apache tribesmen. Forepaugh himself was said to have quelled the trouble by assuring them this was not the case, but the anecdote served to reinforce white opinions about Native susceptibility to superstition.[11]

On July 5 in Fall River, Massachusetts, an elephant nearly killed his keeper after someone fed him a chaw of tobacco. The same day a lion trainer was mauled on his right arm from elbow to knuckle while packing up a cage. En route to Newport, Rhode Island, on July 6, a section of one of the circus trains derailed and overturned a car carrying horses. Two men were seriously injured, one horse was killed outright, and three others needed to be destroyed because of their injuries. On September

26 in Dubois, Pennsylvania, elephant trainer John Poggy was attacked while trying to get the elephant Tip out of its car. Poggy was struck by the elephant's trunk, gored through the abdomen by its tusks, and trampled to death. Once again, blame was placed on local boys having fed the animal apples laced with tobacco and pepper. On October 1 in Pittsburgh, two canvas workers, one white, the other black, were run over and killed while loading a train.[12]

During Forepaugh's 1888 season, the young cowboy S. F. Cody worked with both Doc Carver and Mexis, both masters of the mirror shot.[13] Cody learned most of his trick shooting that season, and incorporated many of the shots into his pistol act. He eventually developed his own variant of the mirror shot using a revolver rather than a rifle: he stood with arms outstretched, the pistol pointing in one direction while he faced the opposite direction, holding a mirror.

Samuel F. Cody finished the season with Forepaugh's circus when it closed at Altoona, Pennsylvania, on October 6. Adam Forepaugh returned to Philadelphia, preoccupied with his $10,000 in wagers that incumbent Grover Cleveland would be reelected as president over Benjamin Harrison. A month later, Cleveland won the popular vote, but the electoral college vote put Harrison into the White House. The election news disgusted Forepaugh, who did not even like Cleveland, but thought him the better bet.

On October 7, 1888, Cody went from the circus's last stop in Altoona straight to Reading, Pennsylvania. Maud Lee had been living there for several months, working as a housecleaner in boarding houses. After a separation of five months, Samuel and Maud were happily reunited, but their joy was short-lived. Three days after arriving in Reading, Samuel Franklin Cody was arrested and thrown in jail for the assault and attempted rape of a ten-year-old schoolgirl.[14]

On the morning after his arrest, Cody was given a hearing in the private office of the mayor of Reading. The reason for the closed-door session was the fear that the large crowd of onlookers who had gathered as word of the sensational charge spread might intimidate the young girls involved. The cowboy gave his name as S. F. Cody and stated his occupation as being a shooter employed most recently with Forepaugh's circus. He explained that he was in Reading to make arrangements to join up with another Wild West show. Cody was described as wearing a corduroy suit over a blue flannel shirt, with nickel trim and red glass stars on his collar. Around his neck he wore a red tie looped and pulled tight with a gold finger ring. He had a broad-brimmed gray wool hat with a leather band, decorated with silver stars. He had foot-long brown hair and a small light mustache.

According to the account of the hearing in the *Reading Eagle,* Mary Resch, age ten, testified to the mayor: "I came down Franklin Street about 8 o'clock in the morning. A man caught me by the foot with a hook tied to a rope that was thrown through a cellar hole as I passed a house above 5th Street. When he caught me, he pulled me down into the alley towards the cellar hole where he was. That is the man," she said, pointing at Cody. "I was quite close to him. I was halfway down the underground alley steps when another man, who was dressed like a fireman, came to me and got me out. I had screamed loudly. The hook at my foot was unloosened when he [Cody] pulled me to him, but I don't know how. When he heard the other man had come to help me, he [Cody] ran out the other way."

The other girl, Gertie Tracy, was just seven years old and had been walking to school with Mary. She was asked what she had seen. She said she had not seen any of the events take place, because she had been walking ahead of Mary. The mayor asked Mary if she had seen Cody before. She answered that she had seen him two days earlier on the street as she was leaving school. Mrs. Resch, Mary's mother, claimed that Mary had told her that the man who had come to Mary's assistance took the rope that Mary's attacker had abandoned and told the girl that he would take it to the police in Philadelphia, where (he said) such articles were known to be used to trap little girls.

Cody was asked what he knew about these charges. "I was not at the house, alley, or cellar that this girl says I was. I did not get out of bed at my boarding house on 200 Franklin Street until after 10:00 A.M. yesterday. I would like to have my witnesses heard. I am innocent of this charge and would not stoop to anything like harming little girls. Last night when the officer took me to see this girl, she said I was the man, but would not look me in the face; and the other girl knew nothing about me at all."

The mayor decided that the legal process had to run its course. Cody was sent to the lockup, with a further hearing scheduled to determine if he could be released on bail. His court hearing would not occur until the second week of December, when the county criminal court would begin its new quarterly session. Cody reacted with obvious alarm: "If you lock me up for any length of time it will kill me! If I have to remain there, I'll kill myself!"

On the way back to the jail, Cody and his lawyer both spoke with reporters. Cody's attorney, J. H. Jacobs, said he was convinced of Cody's innocence. "He comes from that section of the country where a crime of this kind means instant death by lynch law. Cody and his friends here have not the slightest idea why these charges are being made. They can conceive of no motive."

Cody added, "What in the name of Heaven would I mean by trying to assault a little girl? Where is that man who is said to have released her? Why were her cries not heard at the police station just three doors down? Why did the man who helped her not report this dastardly outrage to the police; or bring the rope there? Why has he not come forward? I can prove my innocence, but I hate the charge, and I hate going to jail."[15]

The Berks County Criminal Court convened nearly two months after Cody's arrest, on Monday, December 10, 1888. Cody had to sit in the courtroom several days before his case was called. He placed his broad-brimmed gray hat on the floor next to him. Everyone in court noted that a young admirer sat just behind where Cody was in the prisoner's box, showering her attentions on Cody. During the proceedings, while other cases were called and decided, Cody and the young lady exchanged glances and pleasant words. She would leave on breaks and go outside to get refreshments to bring back to Cody: oranges, candies, and a Pennsylvania Dutch specialty: cinnamon sugar pretzels. Reporters recognized her as having tried to secure Cody's release ever since his arrest. The young woman was Maud Lee, and she was determined to see Cody set free.

Cody's case was called at 4:40 P.M. on Wednesday, December 12. The prosecutor, Assistant District Attorney Schrader, outlined the charges and immediately called Mary Resch to the stand. She appeared in a dark brown suit, with a pink silk cape and a brown, fur-trimmed hood. She repeated her story in a straightforward manner. Gertie Tracy was not present to testify, being at a funeral in Philadelphia. Schrader called no other prosecution witnesses.

Counselor Jacobs, in his cross-examination of Mary Resch, did not suggest that the girl had fabricated the whole incident. Instead, in a fatherly tone, he did ask her to admit it was possible that the missing man—the one dressed like a fireman—might have been the one that caught her with the rope and pulled her toward the cellar. He then addressed the judge with a list of actions that a reasonable rescuer would have taken on seeing a little girl in such a situation—none of which was done by the mystery man. Jacobs asked the court if the Commonwealth should proceed with the case, if the only supporting testimony had already been heard. It was already obvious that Mary Resch's testimony raised many doubts.

Schrader declined to resign the case. Jacobs then called his own witnesses, starting with Cody. Cody again swore that he knew nothing about the girl and that he had not left his room on the day in question until after 10:00 A.M. Under cross-examination, Cody was asked about his skill with a lariat. He admitted to being an expert with a lasso—but a rope with a metal hook at the end was not a cowboy's tool. Moreover, he pointed out that it would be impossible to throw a rope any distance through a cellar hole of the diameter that had been measured.

The managers of the house where Cody boarded, Mr. and Mrs. Mays, attested to the fact that Cody had not left the house that morning and could not have done so without being seen by them. William Bowers, a local store clerk, swore that he saw another stranger dressed as a cowboy walking around town from October 8 through 11, but that it was not Cody. Finally, Maud Lee took the stand and stated positively that Cody did not leave his room until 10:20 A.M. She was certain of the time because she had been waiting to do her job of sweeping out his room.

At that point, the prosecutor, Mr. Schrader, threw in the towel. He had a lengthy conference with Mr. and Mrs. Resch, then addressed the judge:

"While this is a very strange case, your honor, and while we believe the little girl was lassoed as she describes, yet we do not feel like pressing for the conviction of Cody, as he may not have been the man." The judge concurred and told the jury they would be justified in finding a verdict of not guilty. Five minutes later, the jury delivered that verdict and Cody was acquitted and discharged.

Samuel F. Cody exited the courthouse with an expression of obvious relief. Clasping his arm and walking at his side was a triumphant, beaming Maud Lee. The reporters on the scene noted her jubilance as she waved to the crowd outside. She no doubt believed that she had helped save her lover from a horrible fate.[16]

Though S. F. Cody and Maud Lee left the Reading Courthouse victorious, the preceding two months had erased their savings. The Wild West show job that Cody had hoped for in October had vanished, so he had to spread word of his availability to circus and theatrical agents. By luck, a new play was scheduled to open at the New Standard Theatre in Philadelphia less than two weeks after Cody's acquittal. It was a western melodrama that required a large cast of cowboys and Indians. Called *Deadwood Dick; or the Sunbeam of the Sierras,* it starred none other than Annie Oakley.

Deadwood Dick's plot involved a white settler girl taken hostage and raised by Plains Indians, who teach her to use a gun. Annie's cameo shooting demonstrations in the drama might have been lost amid the more than two dozen violent deaths sprinkled through the scenes. The play opened on Christmas Eve 1888—and was soundly panned by critics as dreadful. "The plot is unreasonable and the dialogue is remarkable for its bombastic crudity," said the *Philadelphia Press* the next day. Perhaps the critic had picked up on the fact that the leading man had left town the day before the opening, which necessitated hasty scene cuts and dialogue edits by the unprepared producers.[17]

After a short run in Philadelphia, the production went on tour in January 1889, starting with smaller towns and cities in southeastern Pennsylvania. It may be that Maud Lee accompanied S. F. Cody on the play's tour; years later she recalled having met Annie Oakley during this period. *Deadwood Dick* played to meager audiences wherever it went. In late January, while in Chambersburg, Pennsylvania, it came to a merciful end when the assistant manager, John Keenan, vanished with all the show's receipts.

The cast was stranded; most were owed their pay and could not even afford train tickets or lodging for the night. Annie Oakley and Frank Butler paid for the actors' rail tickets and settled their lodging debts. Most of the cast, including Frank, Annie, and S. F. Cody, headed back to Philadelphia.[18] The cast members initiated a lawsuit against producer John Burke for their back salaries.[19] In February 1889, Annie Oakley and Frank Butler rejoined Buffalo Bill's show, which was about to embark for France to be a feature of the Exposition Universelle de Paris. Lillian Smith's contract with Buffalo Bill had not been renewed, which paved the way for the rapprochement between William F. Cody and Annie.

Adam Forepaugh was ruthless in eliminating troublesome members from his circus. Before the 1889 season opened, Forepaugh sold the lion that had mauled its keeper the previous year to the Chicago Zoo. Tip, the elephant that gored to death its keeper, was magnanimously donated to the city of New York for the Central Park Zoo. Bolivar, the largest elephant in captivity, was literally eating too much of Forepaugh's profits. Since Barnum's Jumbo had died nearly four years earlier, Forepaugh did not think it worthwhile to continue to write off the expense of Bolivar's appetite. With great fanfare, Bolivar was donated to the Philadelphia Zoo.

It is unlikely that the story of S. F. Cody's arrest and trial had escaped the notice of Adam Forepaugh. These events were reported in Philadelphia papers, including the mention that Forepaugh was Cody's most recent employer. Adam Forepaugh would not have been pleased with that type of publicity.

Regardless of whether returning to the Forepaugh show was an option they ever considered, Cody and Maud were already working on a more ambitious plan to develop their own headline act. The public already knew the names of the greatest sharpshooters in the country: Adam Bogardus, Doc Carver, Ira Paine, Annie Oakley, Lillian Smith; but they all used rifles or shotguns. Samuel F. Cody's special talent was shooting .45

caliber Colt revolvers, which set his act apart from the wing shooters. In order to bring Maud into the act as more than just a pretty target holder, Cody had been teaching her trick shooting. Thanks to her father, Maud already had some experience with guns, and she was gifted with a good eye and a steady hand.

Following the collapse of *Deadwood Dick* in mid-February 1889, S. F. Cody and Maud were booked as "Mr. and Mrs. S. F. Cody," in a New York City dime museum, the Star Museum, located on Eighth Avenue between Thirty-Sixth and Thirty-Seventh streets. Maud was still just seventeen years old. The Star Museum had opened in January, joining roughly a dozen other dime museums in Manhattan. Dime museums catered to a lowbrow audience, and the Star Museum was not among the upper echelon of theaters offering similar fare. Not fooled by the Star Museum's claims to educational value, the Women's Christian Temperance Union immediately launched a campaign criticizing the fact that the attraction was open on Sundays.

By the late nineteenth century, dime museums were far removed from their eighteenth-century origins as showcases of libraries and collections of private citizen-scholars. The first museums in America were opened to the public as tokens of belief in the ideals of civic participation and public education. Museums were also seen as means to offer morally instructive popular entertainment as an alternative to immodest, sinful, and degenerate diversions.

The finances of some collectors were outstripped by their intentions, so they began charging admittance to view their treasures. To recoup the costs of maintenance and new acquisitions, as well as to earn a profit, they sought to attract large numbers of visitors through sensational displays. Consequently, live performers were added to the museum premises. The performers that appeared in these museums also found employment in circuses or in legitimate theaters, providing entertainment during scene-changing intermissions. The most notable examples of this combination of encased curiosities and live variety show were Peale's American Museum in Philadelphia and Scudder's American Museum in New York. The format was perfected by Phineas T. Barnum after he purchased Scudder's American Museum in 1841.

By the 1880s and 1890s, there were dozens of museums offering the same mix of remarkable artifacts, wax effigies, panoramic paintings, human freaks, acrobats, dramatic sketches, musical acts, and comedy. Because the typical admittance fee was a dime, they became known as dime museums. The acts at dime museums would change every fifteen minutes or so, and therefore the performers did several shows a day to provide continuous entertainment. During this period, many performers accepted bookings at both dime museums and variety-hall theaters.

"Mr. and Mrs. S. F. Cody" (they were not yet legally married) appeared on the same bill at the Star Museum with an assortment of vaudeville acts.[20] Levando Baldwin, the "armless wonder," demonstrated his ability to eat, write, and paint pictures using his feet. Baldwin was still recovering from the loss of his entire life savings, $250, which had been stolen by his runaway bride two years earlier. At the time, Baldwin was made the butt of a joke published in papers across the country, which suggested that he now wished he had married a "legless wonder."

Another act was Sampson, a native African who was said to have accompanied Henry Stanley on his recent Congo explorations. Samson gave demonstrations of his strength by bending iron bars with his bare hands. He also appeared in tribal dress, spoke in his own language, and performed a Zulu war dance.

Hattie Bell took to the museum's stage as a singer "in the plantation style," meaning that she was a trained minstrel show vocalist, and likely appeared before the audience in blackface and dressed in rags. Acts that reinforced such racist stereotypes were part of the fare of nearly every dime museum, variety hall, and vaudeville theater of the 1880s and 1890s. Ethnic ridicule was not limited to blacks. Also on the bill at the Star were "Big and Little Nick Murphy," who performed comedic sketches as Irish and Italian immigrants. Their act was heavy on dialect, and they told stories that embraced derogatory national characteristics without making their characters unlikable.

Three weeks after their Star Museum engagement, S. F. Cody and Maud appeared at a much larger venue, Worth's Palace Museum. Although it was another dime museum, Worth's Palace Museum occupied five city lots in lower Manhattan. It was operated by George H.

Huber and E. M. Worth; their partnership was short-lived, and a year after the Codys appeared, Huber took over this establishment on Fourteenth Street and ran it as one of New York's most profitable dime museums, Huber's Palace Museum, for twenty more years.

At Worth's the two shootists shared the bill with "Professor Sherman's Caprine Paradox," which consisted of six trained goats. They were drilled to climb ladders, leap hurdles, walk atop rolling barrels and balls, jump through hoops, walk on their hind legs and knees, and balance on a seesaw. Sadly, ten years later, Sherman would lose all his goats after they ate the Christmas decorations hanging in a Philadelphia theater. Apparently laurel wreaths are one of the few organic materials that goats cannot digest.[21]

Also on display was the fakir Ki-Ki, "the human salamander," who was able to walk on red-hot metal plates and broken glass. There was also a chess automaton named Ajeeb, which would engage visitors in live games. This was evidently an imitation of a more famous chess automaton of the same name that was a fixture of the rival Eden Musée. Chess automatons were mechanical figures dressed in eastern Indian garb and perched in a sitting position at a chess table. The pedestal they sat on could be opened to display impressive gears and clockworks, but in reality these were just a façade to a hidden compartment in which a chess master would sit and manipulate Ajeeb's arms. Another act found at Worth's in March 1889 was Laura Howard, an albino who would allow visitors to stare at her flesh, hair, and eyes. Sampson, the Zulu strongman, had moved with the Codys from the Star Museum to also perform at Worth's.

The same week Worth's patrons could also experience the unique sounds of the Guatemalan Indian Marimba Band. Another human freak on display was Mrs. O'Brien, a giantess. The billing included others: George C. Marshall, dancer; Tom Haley, Irish comedian; Annie Wyandotte, dramatic soprano; Harry Bryant (comedian) and Pollie Holmes (contralto vocalist); Dave Reed and family (musical farce); Prof. J. W. Hampton and his educated animals; and Mrs. Connolly, the hairless woman.

During the two months that S. F. Cody and Maud Lee performed in these New York dime museums, they kept an eye on the show-business postings advertised in the leading theatrical newspaper of the time, the weekly *New York Clipper.* It must have seemed like a stroke of good luck when they discovered an ad in the February 16 edition calling for cast members for a new Wild West show. S. F. Cody and Maud Lee responded to the ad, but waited several weeks before they heard back from the new show's management.

In December 1888, Buffalo Bill's main competitor in the Wild West show business, Pawnee Bill (real name Gordon Lillie) had abandoned his unprofitable show—including much of the cast, livestock, and paraphernalia—in Wichita, Kansas. While Lillie looked to sell the whole outfit, Pawnee Bill's stranded cowboys congregated in the yard of the Wichita Horse and Mule Market, a frequent gathering spot for men in search of jobs and conversation.

The proprietor of the Wichita Horse and Mule Market was soon tempted by the idea of entering show business, and with his wealthy friends as investors, he bought out the show from Pawnee Bill. The new owner was not only inexperienced in organizing touring shows, but he also had other matters competing for his attention. For instance, in the early months of 1889, he was also deciding whether or not he would lead a war against the U.S. government.[22]

CHAPTER 4

OKLAHOMA

1889

This prospective showman and revolutionary was "Oklahoma" Harry Hill. By 1889, he was a successful cattle rancher and horse trader. In 1877, Harry Hill had traveled to Fort Worth, Texas, from Missouri as part of his horse-trading business; while in Fort Worth he was inspired by the rhetoric being published in broadsides by Charles C. Carpenter. Carpenter advocated a mass invasion of settlers into the Unassigned Lands section of Indian Territory—what is now the state of Oklahoma. This encroachment was to take place on a scale that would make policing by U.S. federal troops an impossibility.

The Indian Territory had been set aside in 1828 as a dumping ground for relocation of American Indian populations east of the Mississippi. Andrew Jackson's Indian Removal Act of 1830 gave the U.S. president the authority to use the military to accomplish those evacuations. Some tribes did not resist; others were forcibly marched; in the case of the Cherokees, the deadly march became known as the Trail of Tears.

Decades later, the U.S. government used the aftermath of the Civil War as a pretext to renegotiate treaties with the tribes then occupying the Indian Territory. These new treaties ceded rights of way through the territory to the railroads, and reverted ownership of most of the western and central areas of the territory, which belonged to the Creeks and Seminoles, back to the government. The central part of Oklahoma, a wide swath of arable land, was designated as public domain and was known as the "Unassigned Lands."

In the 1870s, public pressure for white settlement of those lands, fueled by voices like Charles C. Carpenter's, began to build. Their legal argument was that the Homestead Act of 1862 already gave a citizen the right to claim a 160-acre lot in the Unassigned Lands; however, there was countering pressure to keep the lands under U.S. government or American Indian tribal control.[1] Both the federal government and the relocated tribes leased Indian Territory rangeland to large-scale cattle companies, many of them operated by Texas concerns.

Carpenter was more adept at propaganda than actual mobilization for this movement, and the primary activist leader was Captain David L. Payne. Payne had been a Union army private in the Civil War. In the late 1860s, he was elected to the Kansas legislature. He also served stints in the Kansas Cavalry, which was raised in response to attacks during the Indian wars of the 1870s—mainly raids by the Cheyenne. During these periods of enlistment he rose to the rank of captain. When not in the legislature or on military duty, he made his living as a hunter, guide, and scout. Payne was never successful at holding a steady job or retaining any land or money, however.

In the late 1870s, Payne went to Washington, D.C., in hopes that his record as a loyal Democrat would gain him a position in the federal government. During those years he was not favored with any responsibility higher than to serve as a congressional messenger and errand boy. Yet that low-level employment did at least afford him the opportunity to overhear the lobbying and floor arguments over the disposition of the Unassigned Lands.

Payne left Washington in 1879 with an empty wallet but inspired with the conviction that his destiny was to lead the settlement of the Unassigned Lands. Upon his arrival back in Kansas, he formed an organization of "boomers," one of whom was Harry Hill. Payne, with Harry Hill as his scout, led an invasion into Indian Territory in May 1880, but they were quickly arrested by the U.S. Army and escorted back to Kansas. Payne tried two more invasions that year, but was again either evicted or stopped cold at the border between Kansas and Indian Territory.

Payne's last attempt to settle a colony took place at a place called Rock Falls in the disputed Cherokee Strip section of Indian Territory. Payne was arrested there by the army in August 1884. The federal government had grown frustrated with both Payne's boomer invasions and the ineffectual legal consequences intended to discourage them, so it gave the arresting officers orders to handle Payne in a different way in order to discourage his persistence.

Instead of being placed on a train to Fort Smith from Caldwell, Kansas—just ten miles away from Rock Falls—Payne and several of his companions were forcibly marched three hundred miles through the prairie wilderness during the worst heat of summer. Payne endured this ordeal, arrived at Fort Smith, and was hailed as a hero by his boomer followers. But a few months later, he collapsed suddenly of heart failure, and the leadership of the boomer movement passed to others: William L. Couch, Pawnee Bill Lillie, and Harry Hill.[2]

Being a fairly successful businessman, Harry Hill was a more temperate voice among the boomer advocates; he had seen the sad consequences for the early would-be settlers of Oklahoma, and by the start of 1889 believed that the federal government would soon acquiesce to the inevitable. While Couch lobbied tirelessly in Washington, the Wichita Board of Trade hired showman Pawnee Bill to lead the Oklahoma colonists. The Board of Trade could perhaps be excused for believing that Pawnee Bill's national celebrity might draw attention to their cause and for overlooking the fact that show-business mavens are prone to use threats and bluster more than patience and diplomacy. As thousands of settlers descended on the Oklahoma border in early 1889, the situation was tense.

Pawnee Bill felt that Congress was not acting quickly enough to pass legislation to open up Oklahoma to settlement, and by February was threatening to invade regardless of federal approval. Couch telegraphed Harry Hill on February 2, 1889, to request that he try to persuade Pawnee Bill to tone down his rhetoric.[3] On February 7, Hill met Pawnee Bill on the border at Caldwell, Kansas, and told him to rein in the anxious boomers.[4] Meanwhile, in Washington a bill was approved in the House of Representatives that opened the Unassigned Lands to settlement—but that bill still needed to pass the Senate.

Meanwhile, on February 25, newspapers reported that Harry Hill had commissioned a circus train to be built, and that it would be used to carry his Wild West show from Wichita to New York City on April 25.[5]

In early March, the U.S. Senate voted down the Oklahoma measure. The growing masses of encamped settlers asked Harry Hill to help lead a defiant incursion. On March 10, Hill told the boomers at Caldwell and Hunnewell that he would join them, and that if the army pursued them, they would cut fence wires and burn grassland to destroy the business of the cattle companies leasing rangelands in Oklahoma.[6] But Harry also sent a wire to newly inaugurated President Benjamin Harrison, urging him to declare the Unassigned Lands open to settlement. On March 16, Hill arrived in a boomer camp already erected within Indian Territory, near the Arkansas border in the Fort Smith area. Soldiers rode out from the fort and forced them back across the border at bayonet point.[7]

A week later, President Harrison proclaimed the Unassigned Lands open and that settlers could establish claims starting on April 22, 1889. The great Oklahoma Land Run of 1889 was on, and a war against the U.S. Army was averted. Harry Hill had less than a month to put together a Wild West show and get it on the road; however, his recent travails provided a new theme for his show: it was to be called the "Oklahoma Historical Wild West Exhibition" (sometimes also labeled the Historical Oklahoma Wild West Exhibition in promotional posters).

As soon as Samuel Franklin Cody and Maud Maria Lee received word from Harry Hill to come to Wichita, they made immediate wedding plans. On April 9, 1889, they arrived at the Montgomery County Orphans' Court in Norristown, Pennsylvania, in the company of Maud's parents, Joseph and Phoebe Lee. They took out a marriage license, and five minutes later were pronounced man and wife. Samuel was now twenty-two; Maud had just turned eighteen. Maud's father told the papers that the pair had become "kinder thick" since meeting a year earlier. The *Philadelphia Inquirer* pronounced the marriage a romantic sequel to December's Reading trial. The new Mr. and Mrs. Cody then caught a train headed west.[8]

Even as the Codys were on their way to Kansas, Harry Hill started to spread publicity linking his show's celebrities with the Oklahoma Land Run. With or without the performers' knowledge, Hill announced that several of his stars would take part in the April 22 Land Run and claim their own 160-acre quarter-sections: Adam Bogardus and sons (Bogardus now called Kansas his home); Texas Ranger "Buckskin Roy"; the show's announcer Captain Harry Horne; "Towanda Charley," a cowboy from the Black Hills of Dakota; and "S. F. Cody Jr. and wife, the great pistol shots of the Pacific slope." It was explained that all would file claims, then return from the tour within six months in order to legally prove their claims. Hill's press release stated: "Mr. Cody is a cousin of W. F. Cody (Buffalo Bill), and a much finer shot. His wife is also an expert, and, should they succeed in getting a claim marked out, it would be a sorry day for the claim jumper that attempted to beat them out of their rights. One of Cody's tricks is to throw into the air a 10 cent piece and then with a pistol ball mark it."[9]

Harry Hill was in a position to make sure his stars made the best of this opportunity, if they so desired. He knew the Unassigned Lands as well as any white man alive, and would have been able to direct them to the best tracts. Some lots were the equal of the Promised Land that all the Oklahoma settlers dreamed of: clear creeks with clay banks, large stands of trees, good soil, and abundant game and fish. Hill knew the best routes to take and how to ford the rivers flooded with spring runoff; and he had a supply of good horses from his livery business.

Tens of thousands of settlers gathered on the Oklahoma border on April 22—and Sooners had already entered the territory in advance of the official starting hour. Many were poor and desperate, and gambled that Oklahoma would be their salvation. Many were ill equipped and unskilled as homesteaders, and even an unschooled eye could tell their efforts were doomed to fail. Some spent more of their fortunes getting to Oklahoma than the lots were worth. They wasted their money on weak livestock, or lost it to the gamblers, thieves, and con artists that swarmed into the settler camps.

More than a few would strand themselves short of their goal after riding their mounts to exhaustion and death. According to one estimate, there were an average of six claimants contesting each lot. Even so, it would have been hard for a witness to the border spectacle not to get caught up in the feelings of hope and optimism, and the idyllic vision that Oklahoma represented. Samuel F. Cody had all the necessary skills to be a successful homesteader on the plains. For him, it would be a return to the landscape and way of life that he had mastered as a teen.

It is not known whether Samuel and Maud rode among the settler tent camps, dashed off with the crowd of horses and wagons when the shots sounded at noon on the twenty-second, or reached a lot that could have been their own personal Eden. Their own talents (and those of Harry Hill) promised every likelihood of success if that had been their choice, but they never filed a claim. Instead, fate and ambition led them far from the homesteading life that Oklahoma offered.

The Oklahoma Historical Wild West Exhibition left Wichita for Emporia, Kansas, on May 5, 1889, in Harry Hill's seven-car custom rail train. Onboard were seventy-five cowboys, vaqueros, sharpshooters, cowgirls, and property hands; fifty Pottawatomie Natives; fifteen buffalo; and nearly one hundred ponies, horses, mules, and steers. Its stars, besides Bogardus and the Codys, included Captain Harry Horne, Yellowstone Vic Smith, Don Ze Anno, Malo Espanosta, Buffalo Bessie, the "Wichita Kid," California Frank, Texas Mac, Bronco Charley, Buckskin Roy, and Texas Charley.[10] To draw crowds, Professor T. F. Lanford made parachute leaps from a small hot-air balloon—a dangerous daredevil stunt that over the next twenty years would become a staple of state fairs, amusement parks, and circuses.

As circuses had long done, the Oklahoma Historical Wild West Exhibition announced its arrival at a new engagement with a parade. Harry Hill's parade was led by a bandwagon drawn by six white mules, followed by two old stagecoaches, the American Indians in their traditional dress, the cowboys, vaqueros, Spanish lady riders, and settlers. The show opened with Captain Harry Horne introducing each of the stars. The first act was a shooting demonstration by Adam Bogardus and his sons.

Another shooting act was thirteen-year-old Harry Hill, Jr., the manager's son. Further into the program the cowboys, led by the masterful Yellowstone Vic, would race their horses and—at full speed—dip from their saddles to pick up a handkerchief from the ground; Samuel Cody might have joined them on occasion, since he was known for this trick back in his Montana days. The cowboys would then try to stay upright on bucking broncos as they careened around the arena.

As the show progressed, the Codys were brought out for their pistol-shooting act, breaking dozens of target balls tossed into the air, held in each other's hands, or dangled from their mouths. Harry Hill, Jr., made a reprise appearance, shooting a small glass plate held between Maud's fingers—Maud obviously had a rapport with adolescent Harry Hill, and he undoubtedly treasured the confidence that the young woman showed in his aim.[11] The Mexican equestriennes, led by Malo, raced against other cowgirls for the prize of a gold watch.

After these exhibitions, the show presented a series of historical scenes: the first was the capture of wagon master Pat Hennessy by Cheyenne and Comanche Indians, who tied him to a wagon wheel and threatened him with burning, before his last-minute rescue by a passing group of cowboys. This scene was based on an event from 1874 that still stirs controversy. In reality Hennessy and his group were massacred; today, there is some doubt whether Indians were to blame.

This was followed by a lighthearted sketch in which a father forces would-be Romeos on horseback to chase after his daughter on her fiery steed, with the winner to claim her hand. Then the Indians performed a wedding feast for their chief and a war dance; a horse thief was lynched; and a stagecoach (driven by Harry Hill) was held up by highwaymen. Finally, no show was complete without a depiction of the arrest, tortuous march, and imprisonment of Harry Hill's old friend, poker companion, and "Father of Oklahoma," Captain David Payne.[12]

From Emporia the Oklahoma Historical Wild West Exhibition headed to Kansas City, Missouri, where it competed for ticket sales with an appearance by Captain Paul Boyton's Aquatic Exhibition.[13] Next it headed as far north as Milwaukee. By the end of May the show traveled south to Chicago, then on to major dates in Louisville, Kentucky, Cincinnati, Ohio, and Springfield, Illinois. Harry Hill's original ambition was to take the show from the East Coast to the West Coast, but by August 3 it limped back to Wichita after substantial losses and many missed dates due to bad weather. Like all traveling shows, it suffered its share of mishaps, but no apparent fatalities.

After performances for his hometown audience, Harry Hill had to abandon plans to move the Wild West show on to California. His show-business dreams ended where they had begun less than four months earlier.[14] The Codys' headline billing was short-lived. They must have had doubts, if only momentarily, about whether their ambitions were any more substantial than the Promised Land dreams of the Oklahoma Boomers and Sooners.

CHAPTER 5

GREAT BRITAIN

1890–1891

The failure of Harry Hill's Oklahoma Historical Wild West Exhibition in August 1889 left S. F. Cody and his young wife stranded in Wichita. Their actions during the following eight months are not well documented. In later years, Maud mentioned performing in Miles Orton's Circus and Wild West—a show that toured only in 1889.[1] So it is possible that the Codys salvaged the touring season by joining Orton's Iowa-based aggregation for the remainder of the summer.

"Samuel J. [*sic*] Cowdery, Jr.," was listed in the Davenport, Iowa, City Directory for 1890–91, which was published early in 1890. He was living at the same address as his mother, Phoebe Cowdery. While it might be understandable that S. F. Cody wanted to introduce his new wife to his mother, and that he and Maud might have completed their tour season in the Midwest, the extended period in which they were out of the public eye seems curious. Ambitious entertainers usually do not spend half a year in idleness.

In January 1890, the *New York Clipper* printed an advertisement by a small, well-established circus that was looking to expand by offering Wild West acts.[2] The Codys answered Washburn and Arlington's "United Monster Show" casting call, and in the spring headed to the circus headquarters in Bristol, Pennsylvania. The circus boasted that it had one hundred gray horses and one hundred performers, but it also had only one ring under its tent. Washburn and Arlington's 1890 season route headed north through New Jersey, past New York City, and into Connecticut. By Saturday, May 10, 1890, the circus had reached Stamford, Connecticut, where it opened its afternoon show at 2:00 P.M.

During that performance, S. F. Cody and Maud performed their pistol-shooting stunts, dazzling the audience with their mastery of .45 caliber Colt revolvers. Then Maud picked up a rifle and it was announced that she would shoot a half-dollar coin while aiming over her shoulder using a looking glass—in other words, the mirror shot. No other reviews of their act up to this performance mentioned their use of this stunt in their act, nor even any use of rifles.

As Maud leveled the rifle on her shoulder, strands of her long brunette hair strayed over the top of the barrel. As she moved to brush her hair back, her hand knocked the barrel of the gun. The movement forced her finger against the trigger and the rifle fired.

Cries rose from the audience, followed by screams. A young girl sitting in the crowd about twenty feet away from Maud had been hit by the bullet just above her right breast. Mary E. King, age twelve, slumped in her seat as people rushed to her aid. Maud stood frozen in horror.

The bullet had entered the girl's chest without exiting; at first the attending doctors feared it was close to her heart. They were reluctant to try an extraction and sought an experienced surgeon. Mary's condition was described as critical, and her family was warned that her prospects for survival were not good. Maud was placed under house arrest at her hotel. Plainly shaken, Maud told reporters that she would never touch a rifle again. The circus moved on from Stamford to Norwalk, but the local sheriff attached it with an order restraining the show from opening in or moving from Norwalk until Maud's hearing.

The next day, a surgeon was able to extract the .22 bullet from beneath Mary King's armpit, and her chances of recovery were said to be improved. Maud was brought before a judge who heard the circumstances of the accident; he was assured by witnesses that the gun had not been intentionally pointed in the girl's direction. The shooting was ruled accidental, and Maud was released.[3]

It is impossible to tell what impact the girl's shooting had on nineteen-year-old Maud Cody. Although Maud swore to reporters that she would not touch a rifle again, she cannily did not mention revolvers. She may have hoped that their tour would go on, but Washburn and

Arlington evidently decided that they could not afford a similar mishap. The Codys did not move on with the circus.

S. F. Cody's reaction to their career impasse was to suggest that they could get a fresh start by going abroad. They had experienced a string of bad luck: Cody's arrest in 1888, the collapse of the Oakley stage production, the stranding of Harry Hill's show, the prolonged inactivity in Davenport, and Maud's accident in Stamford. Because Buffalo Bill Cody's greatest success had come overseas in England in 1887, S. F. Cody reasoned that their fortunes might fare better across the Atlantic. Without any advance bookings, and with Maud's psyche in fragile shape, he decided that he should scout the territory by himself.

S. F. Cody left the United States on June 10, 1890, less than a month after Maud's mishap. When Cody disembarked in Liverpool he likely had little money and no contacts. He would have found the British theatrical landscape a bit different than that in the United States. Although the variety format was prevalent on both sides of the Atlantic, in Britain it hailed from a tradition more steeped in musical comedy.

British music halls were an outgrowth of eighteenth-century coffeehouses and taverns where men gathered to share song and cheer—and often quantities of alcohol. These assemblies soon evolved into "song and supper" clubs for the upper class and "harmonic meetings" in working-class taverns. The Theatre Act of 1843 forced saloons either to become full-fledged, licensed theaters—without alcohol—or to retain their drinking licenses but not stage full dramatic productions. It took only a short time for tavern keepers to realize that the Theatre Act allowed them to present musical sketches and other entertainments in a variety format. By the 1850s and 1860s, buildings were renovated or built specifically to stage live variety shows. Many of these new structures had "Music Hall" in their names.

Although many performers traveled from America to Britain or vice versa and performed the same basic act, in general there were fewer billings on British music-hall stages of freaks, acrobats, animal acts, or other performances derived from the circus tradition. In England, such acts were not unknown to British music halls, but were usually relegated to dime museums, which were known in England as "panopticons."

S. F. Cody, with little money and an unimpressive résumé, stepped ashore in England determined to break into the world of the British stage. The best assets he had were his exotic, handsome looks; his laconic western charm; and the helplessness of an obvious foreigner. Perhaps this explains why the first two billings he found were in theaters managed by women known for their soft hearts.

At St. Helens, a town just outside of Liverpool, Cody was booked at the Grand Circus and Theater, managed by Miss Minnie Gough. Miss Gough had just demonstrated her charity by giving half of one day's gate receipts to striking workers at a nearby plate-glass factory. She placed S. F. Cody on the bill with Sandy and Carl, a duo of "entertaining negroists" (who apparently were real black comics, not whites in burnt cork blackface); Miss Edie Morton, a serio-comic; song-and-dance men Will Edwards and Lou Merritt; vocalist J. Murray; Irish songstress Eva Johnston; and F. Carlin, an Irish comedian and dancer.[4]

A week later, Cody traveled north and found another sympathetic theater manager, Mrs. J. S. Baylis of the Scotia Music Hall in Glasgow, Scotland. Mrs. Baylis also had a reputation for being generous to the poor and honest in her dealings, and was known to accept any acts from America that came her way. As soon as Cody was offered a contract, he sent word across the Atlantic for Maud to join him. Mrs. Baylis not only gave Cody a three-week engagement, she also supported his efforts to get additional bookings at other theaters by placing large ads in the *London Era,* Britain's theatrical newspaper, giving his contact address as the Scotia Music Hall.

At the Scotia, Cody joined several big names on the billing. The headliner was the celebrated humorist W. F. Frame, the most popular Scottish comedian of his era. Frame was best known for his parody of Sir Walter Scott's "Macgregors' Gathering." Professor Bernard Sloman, the "Man Bird," warbled and whistled in imitation of various fowl. Sloman had been perfecting his act for more than sixteen years. Rounding out the roster of performers were comic vocalists George Byford and James Fawn, each a solo act.[5]

By July 26, 1890, S. F. Cody's name was joined on the bill with "Miss L. Cody." Cody had persuaded the newly arrived Maud to take a stage name, and he preferred Lillian. That name evoked two of the most famous beauties of contemporary Anglo-American theater: Lillie Langtry and Lillian Russell. Moreover, Cody's little sister was nicknamed Lillie, and Lillian Smith had been one of Buffalo Bill's famous sharpshooting stars. For the rest of her show-business career, Maud went by the professional name of Lillian Cody.

From this point forward Maud was never billed as Mrs. Cody, but was instead identified as Samuel's sister. It was more usual than not for husband-and-wife acts to maintain separate identities, so the change in their preferred billing cannot be construed as evidence of a change in their relationship. More likely, Mrs. Baylis or some other person seasoned in the show-business world advised them that single performers appealed more to audiences than did married couples.

By the next month, September 1890, Cody and Maud found an engagement farther south, in Birmingham, England, at the Gaiety Concert Hall. The headliner was an eccentric fixture of British music halls, George H. Chirgwin. The middle-aged Chirgwin had been raised in a theatrical family and had been performing in blackface minstrel shows since his youth. As he refined his singing and comedy act over the years, he modified his costume into a weird combination of minstrel and harlequin: he donned a top hat more than two feet high, painted a white diamond shape around one eye of his blackface, and wore a tight black suit. He was known as the "White Eyed Kaffir."

Joining the Codys and Chirgwin on the list of performers were Michael Nolan, an Irish character comic; Maggie Day, a male impersonator; Lloyd Townrow, a noted light sketch actor; and Arthur Ellis, another Irish comic. In retrospect, one has to wonder at how many witty ditties sung in brogue audiences were expected to tolerate. A long-haired cowboy and cowgirl shooting in each other's direction with daring precision must have been a refreshing change of pace.[6]

The Codys were hardly the only American-style shooting act touring British music halls, however. Frank Butler's old partner (and shooting victim), Samuel Hyde Baughman, had been touring Britain with his wife, Lizzie Aldine, since 1887. Other male-female shooting acts found in provincial theaters in 1890 were the Manard Team, Captain James and Lillie Webb, Miami and Texas Charlie, the Levardos, and Colonel and Mlle. De Rizio.[7] There seemed to be more shooting acts in Britain, drawn by the success of Buffalo Bill's first tour of 1887, than there were in America.

Reaching London in mid-September, S. F. Cody and Maud searched for engagements. Though they met with hard luck at the variety halls, fate led them to the grounds of the huge Earls Court exhibition center. Earls Court was a tract of land located next to several rail lines in West London. Several years earlier, an entrepreneur from Yorkshire named John Robinson Whitley had had the vision of transforming the largely disused acreage and cavernous railway sheds into an exhibition center. The sheds were remodeled to provide indoor galleries, and a large arena and grandstands were built adjacent to the sheds for outdoor events.

Whitley's vision for the venue was inspired by the success of several world's fairs during the latter half of the nineteenth century. Earls Court was not large enough to host a full-scale world's fair, so Whitley's approach was to sponsor a series of "national exhibitions" that would concentrate on a single nation each year. From each nation he invited manufacturers, craftsmen, artisans, and makers of agricultural goods to display their wares inside the halls, while in the outside arena he arranged entertainment spectacles along national themes.

The first of these national exhibitions took place in the summer of 1887, when Earls Court hosted the American Exhibition. However impressive the artworks and goods were inside the galleries, the sensation at the American Exhibition was the arena show: Buffalo Bill's Wild West. It was the first trip abroad for William F. Cody's troupe. The modest success the show had in America was transformed into a worldwide sensation by its stay in London, where it was embraced by fashionable British society.

The next year, 1888, Whitley produced an Italian Exhibition. Inside the Earls Court buildings, the Italian Exhibition featured painting galleries, photographic exhibits, gardens and grottos, statuary, a re-creation of the Roman Forum, and a Pompeii street scene. On the outside grounds, an alpine switchback railway (that is, a scenic rollercoaster), ran through the twenty-four-acre site. The open-air amphitheater was made into a full-size Coliseum of Rome with an arena show featuring Caesar's procession, wrestling, footraces, gladiators, and chariot races. The Italian Exhibition was still running in late summer, the same period when Jack the Ripper terrorized London. Some observers questioned the wisdom of re-creating bloodthirsty Coliseum scenes while a real killer roamed the city streets.

Whitley took a break from his national exhibitions in 1889, the year that Paris held the Exposition Universelle, one of the greatest of all world's fairs. The Paris fair was a huge attraction, drawing upwards of 30 million visitors. Its memorable features included the Eiffel Tower, vendors and entertainers from North Africa, and another triumphal engagement of Buffalo Bill's Wild West. One audience segment that the Paris fair did not attract was the royal families of Europe, who were reluctant to celebrate the centennial of the mass bloodletting of the French Revolution.

Taking note of the popularity of the Exposition Universelle, John R. Whitley realized that many potential middle-class visitors from England were not able to travel to France. Therefore he brought the Paris fair to them with his 1890 Earls Court show, the French Exhibition. Many of the displays were directly imported from the previous year's Exposition Universelle, and a noted French scenery painter, Marcel Jambon, was hired to re-create French panoramas on acres of canvas panels. A scale replica of the Eiffel Tower was erected to serve as the exhibition centerpiece. Teams from France oversaw the construction of an orchid greenhouse, a Norman cider mill, cafes, and a restaurant. A French company was hired to illuminate the grounds with electric light, a recent invention. Inside one Earls Court building, a theater and a circus ran a continuous variety show of French performers: marionettes, a magician, trained animals, military and concert bands, dancers, singers, a hypnotist, and even a séance.[8]

London audiences had already experienced the fabulous arena show that headlined the Exposition Universelle—Buffalo Bill's Wild West— so Whitley decided to retain the French theme with an arena show called the "Wild East." The Wild East, like Wild West shows, presented scenes from lands where white Europeans were attempting to subdue and civilize supposedly wild and barbaric peoples.

The setting of the "Wild East" was France's North African colonies, Algeria and Tunisia. Instead of American Indians, the noble savages of the Wild East were Arabian horsemen, camel drivers, and dancing girls from Biskra; Tuareg warriors of the Sahara (an ethnic mix of Berbers and blacks); shepherds from Constantine; and wizards and musicians from Setif. A total of ninety of these representatives were imported to a temporary village at Earls Court. They were led by a tribal leader, Sheikh Larbi Ben Kess-Kess, a French ally who hailed from an area on the edge of the Sahara between Morocco and Algeria famous for its conical, fossil-rich hills.

Arriving in England after a terrifying ocean voyage, the tribesmen set up a tent village on the Earls Court grounds in the same plot of land where Buffalo Bill's cowboys and American Indians had camped. Once settled, they began to set up displays of sweetmeat cooking, pipe making, scimitar forging, burnoose weaving, dye making, snake charming, primitive dentistry, and basket weaving.[9] Exhibition visitors were invited to stroll through the tent city.

Inside the Earls Court amphitheater, the Wild East's daily program began with the entrance of Larbi Ben Kess-Kess and his horsemen, followed by men in the uniform of the Chasseurs d'Afrique, the noted French colonial cavalry regiment. The "chasseurs" of the show were not French, they were British actors. After all, the French army was not about to send one of its crack desert units to the middle of the damp English capital for six months. After the entrance, Larbi Ben Kess-Kess commanded his riders to show off their saddle skills. Then the sheik presided over a mock Arabian wedding, with dancing and martial sports, including a tug-of-war.

Next on the program were demonstrations by a fire walker, an acrobat, a snake charmer, and a scimitar swordsman. Then a series of races took place between the expert Arab horsemen and the chasseurs. The arena was transformed into Saharan dunes, with sunlight filtered through an orange cloth canopy to simulate the burning sun. A marauder's caravan of camels paraded around the ring. It was attacked by the charging chasseurs with their rifles blazing. A thankful tribal leader (Larbi Ben

Kess-Kess, again) then challenged an officer of the chasseurs to a horse race—and beat him easily, thanks to his iron-gray Arabian steed.[10]

When the Wild East first opened in late May 1890, it proved to be a popular attraction; but it was judged as being not up to the thrilling standard set by Buffalo Bill three years earlier. The Wild West's dramatic scenes had been set to music played by a cowboy band, but the Wild East relied on North African musicians. English audiences described the foreign music as monotonous.

The show's snake charmer was advised to drop one repulsive stunt from his act: swallowing a yard's length of reptile head, then retracting it.[11] This sight did not sit well with audiences, especially those who had just enjoyed a rich repast at the nearby gourmet French restaurant. The scenes with the chasseurs and Arab warriors were also criticized, in that the mounted men dashed after each other shooting rifles into the air, but never seemed to hit anything. Buffalo Bill's sharpshooters had taught audiences to expect skillful gunplay, not just loud bangs.

The initial lukewarm reviews of the Wild East may not have concerned Whitley. During 1887's American Exhibition, crowds flocked to see Buffalo Bill's show and ignored the rest of the exhibits, to the frustration of the vendors inside the cavernous Earls Court halls. To avoid a repeat of disgruntled sponsors, Whitley opened the French Exhibition in mid-May and delayed the first performance of the Wild East for two weeks. He also hinted at sensational acts that would be joining the Wild East over the course of the summer: the lion tamer Darling would harness three of his big cats to pull a chariot; and the King of Dahomey would appear with a bevy of Amazon bodyguards. Whitley also later added a sensational attraction inside one of the halls: a huge aquarium in which deep-sea divers reenacted the underwater scenes from Jules Verne's *Twenty Thousand Leagues under the Sea.*[12]

By September 1890, Whitley addressed the reviewers who had called the Wild East program "aimless" by strengthening the narrative interplay between the chasseurs and the Arabs. Added to the bivouac of the chasseurs were their *cantinières,* women who were contracted by the French military to provide extra food, drink, and nursing, and to serve as mascots in parades. Often these women were the wives of men in the unit they served, and would remarry a comrade in the same unit if their husband was killed. In a cavalry unit like the chasseurs, the women also had their own horses. The Wild East show added mounted cantinières to the opening parade, where they showed off their fancy riding skills in a manner similar to the cowgirls of Buffalo Bill's or Adam Forepaugh's shows.

The retooled Wild East show also introduced a skit about a traitorous French soldier, his rightful punishment, and a fight for the standard that had been wrested from the troops. In other words, more melodrama and, thanks to the cantinières, more sex appeal were injected into the program. Fortunately for the show's producers, there were two performers in London at leisure to provide just what the Wild East needed: Samuel F. Cody and Maud Lee. Cody joined the cast as an officer in the chasseurs, Lieutenant de Franceville. Maud took the part of Mlle. Louisette, one of the cantinières.[13]

The Codys spent two months in the fall of 1890 acting in the Wild East show before the French Exhibition closed in the first week of November. Their steady engagement at Earls Court also created the opportunity for their next billing. Earls Court was in the Hammersmith section of West London, located close to a recently constructed exhibition building, the Olympia National Agricultural Hall. S. F. Cody and Maud must have been encouraged to learn that a Wild West–themed show was being planned at Olympia. As it turned out, though, their next engagement was even more bizarre than the Wild East one.

While it was true that in this new show they would be demonstrating their shooting skills while dressed in western garb, the setting was unlike anything they had ever encountered before. In fact, it involved three elements of Victorian popular culture that survive today only in much-changed forms, making the setting seem even more obscure to anyone trying to imagine it more than a century later. In order to appreciate what S. F. Cody and Maud Lee experienced in December 1890, one needs to understand three uniquely Victorian phenomena: cast-iron–and–glass architecture, stage burlesques, and the roller-skating craze.

The mid-1880s was the heyday of cast-iron–and–glass construction, made possible by innovations that came earlier in the century in producing strong, flexible wrought-iron glazing bars. The first builders to realize the potential of these materials were not architects but engineers of greenhouses and conservatories. Greenhouses were needed to cultivate exotic plants brought to Britain from its far-flung empire, and soon became a standard element of the grounds of country estates in order to support the gardening tastes of the aristocracy.

Horticulturists, mainly John Claudius Loudon and Joseph Paxton, experimented with greenhouse roof design and crisscrossed iron trusses; then applied their knowledge to non-greenhouse structures. The main benefit of cast-iron–and–glass construction was its ability to enclose a large open area with few internal supporting columns and without heavy weight-bearing walls. This made it an ideal building technique for greenhouses, train sheds, grandstands . . . and exhibition halls.

Exhibition halls came into vogue to host agricultural shows, industrial/trade shows, and entertainment spectacles that were too large for conventional theaters. Commercial shows blossomed in the late 1800s, fueled by the increased foreign trade made possible by telegraphy, railroads, and steamships, and by the need to promote all the new technological inventions of the Industrial Revolution. The most famous British exhibition hall was the Crystal Palace, designed by Joseph Paxton and built to house the Great Exhibition of 1851. Many called the Crystal Palace an engineering marvel, but architects and artisans were less impressed, some of them calling it an aesthetic monstrosity. The same reaction was provoked many years later by another iron structure built by an engineer for an exhibition—Gustave Eiffel's tower.

At any rate, the Crystal Palace was a popular success, and spurred the construction of more gargantuan exhibition and entertainment halls in London: Alexandra Palace, the Islington Royal Agricultural Hall, the Albert Palace in Battersea, and the Olympia National Agricultural Hall. Olympia, completed in 1886, boasted the largest roofed arena in Europe. Blueprints also showed a private salon attached to the hall labeled the "Prince's Apartments," said to have been made for Prince Edward for his romantic liaisons. By the time Olympia was completed, there were more exhibition halls in the London area than there were shows to stage inside them. Olympia's builders had hoped to lure the popular simulated battles of the Grand Military Tournament and Assault-at-Arms from the Islington hall. That hope never materialized, but Olympia opened with perhaps an even stronger event, the Hippodrome Circus from Paris, the largest circus in Europe.

The Hippodrome Circus played at Olympia in both 1886 and 1888, but in the intervening years, the hall struggled to make money with sportsmen's exhibitions, sporting dog shows, horse shows, and an Irish exhibition. In 1889–90, P. T. Barnum's Circus took over the Olympia and briefly returned it to solvency. By the spring of 1890, however, it was clear that a handful of limited engagements, no matter how popular, could not make Olympia profitable on an annual basis. Olympia's management decided that what the hall needed was a year-round attraction. They decided to partner with an America promoter to build the world's largest roller-skating rink inside Olympia's arena.[14]

To understand why Olympia's owners gambled the financial future of their magnificent structure on roller skating, one needs to appreciate the immense popularity of the skating fad—a fad that had already seen several ups and downs in the decades before 1890. The public was first exposed to roller skates in the early 1800s when they were used in theatrical productions to simulate ice-skating scenes. Large public rinks were opened in London in the late 1850s. After a burst of popularity, rink attendance waned until a new, improved skate design, the quad, was brought on the market by American James L. Plimpton. Plimpton's design allowed the paired wheels to pivot, greatly enhancing the skater's ability to turn and stop. Plimpton's design gave a boost to the sport in the 1860s. A decade later, ball-bearing wheels and toe-stops were added to skates, resulting in a basic design that endured for a hundred years.

The Olympia rink promoters pinned their hopes on a new refinement in skate design—the use of hemacite-coated wheels. Hemacite was a composite based on an ancient mud-brick formula made from cattle blood and sawdust, then subjected to intense pressure. The result was a material similar to Bakelite plastic that offered a smoother ride than wooden or metal skate wheels.

Olympia's arena was decked with an acre and a half of mitered beech-and-maple paneled flooring. The Olympian Club Elite Roller Skating

Rink opened in April 1890, heralded by the fifty-member Honourable Artillery Company Band. Over the next months, the rink managers held special events to help market the new facility. They sponsored competitive costume balls, fancy skating exhibitions, trick skating demonstrations, and skating races. By late 1890, the Olympian Club was staging theatrical productions on skates to attract its patrons. One of the most popular forms of theater at that time was "burlesque," a type of light slapstick comedy that satirized other dramatic productions, using witticisms, gender reversals, and mildly risqué material to parody the original works.

In the early decades of the twentieth century, "burlesque" became synonymous with blatant sex shows; namely, the lowbrow striptease act. In the latter part of the nineteenth century, however, theatrical burlesques were presented at mainstream venues. A common element of Victorian burlesque was to replace men's roles with casts of women, scandalously underdressed in pants or tights. The humor relied heavily on puns and wordplay; and the music was often pastiches of popular songs. The rarely published scripts were revised frequently to insert topical references to current events. Sly references found in burlesques were likely aimed at young men-about-town and their companions. The audiences that frequented burlesques were therefore urban and upper class. Burlesque was daringly subversive, and it would have appealed to the same crowds that patronized roller-skating rinks.

The Olympian Club management hired Frank Hall, a producer of burlesques, to stage a series of shows in 1890 and 1891. In picking an object to lampoon, Hall selected Buffalo Bill's Wild West, which had created a sensation in London in 1887. Even burlesque needs to bear some semblance to the original, however, so Hall was undoubtedly greatly interested to hear that a real American cowboy sharpshooter was appearing at nearby Earls Court—and that he bore an uncanny resemblance to Buffalo Bill. Moreover, he shared the Cody name and was accompanied by a woman who could shoot like Annie Oakley. So with little delay Hall cast S. F. Cody in the Buffalo Bill role, and Maud was cast as "Any O'Klay."[15]

S. F. Cody and Maud were booked for Hall's burlesque "Wild West," which opened on December 26, 1890 (Boxing Day), and continued for a two-week run. On opening night, the audience inside the hall numbered between seven and eight thousand. The rink floor was open for skating during the day. Other events preceded the Wild West burlesque, including roller-skating races, figure skating, and skater-versus-bicyclist races.

It is not known whether Samuel F. Cody and Maud Lee were forced to sing or strap on skates during the burlesque, but the rest of the cast was, including the pantomime mustangs. It is not hard to imagine that S. F. Cody and Maud Lee experienced a moment of self-awareness in the midst of this extravaganza, as they stared up at the largest ceiling in the world and waltzed among leggy, skating cowgirls; fake American Indians singing ditties; and pantomime horses. They must have felt that they had come very far from the gritty world of cattle drives on the lonesome plains of the American West and the foundry-worker abodes of Norristown, Pennsylvania. They must have exchanged looks, if not words, that expressed, "Can you imagine anything more outrageous than this?"

The Codys were held over at Olympia for another month, until mid-February. They may have returned to music halls in March, but over the Easter holidays in April 1891, the Wild West burlesque was revived at Olympia for another two weeks. Though successful, this run of the "Wild West" burlesque might have been shortened by a legal writ issued against the show's producers by English representatives of W. F. "Buffalo Bill" Cody.

Buffalo Bill was protective of the term "Wild West" and had already fought legal battles against Doc Carver, Adam Forepaugh, and others who attempted to use it. W. F. Cody's English agents were aggressive on his behalf, since Buffalo Bill's Wild West had already scheduled a return to London in the following year, 1892.

Following their Olympia stint, S. F. Cody and Maud were placed on the program for an outdoor variety event held at the Half Moon Grounds in Putney, London, scheduled for the holiday of Whit Monday, May 18, 1891. The advertising for the event promised "representations of a portion of their great drama, *The Wild West.*" Whether this was a wry reference to the Olympia burlesque or an allusion to a script that the Codys were creating on their own is not clear, but at any rate what they performed was their shooting act. The printed program also claimed that S. F. Cody was the son of Buffalo Bill.

Another performer on the program at the Putney show was a parachutist who used the name "Professor Charles Baldwin."[16] Charles Baldwin had made his reputation as a famed aeronaut by trading on the name of Captain Thomas Baldwin, the American aeronaut who had created a sensation with his parachuting demonstration in England several years earlier. S. F. Cody and Charles Baldwin were hardly alone in their appropriation of famous names—popular culture in the late nineteenth century was awash with imitators.

The Putney program's printed reference to "Wild West" is evidence that the Codys did not take seriously Buffalo Bill's earlier legal threats. The show's stage manager, Frank Albert Hughes (who was known in show business as "Frank Albert"), was taken to court for using the term "Wild West."[17] In court, Hughes was asked for the Codys' current address, which he was not able to provide. The case dragged on through the summer and fall of 1891. Hughes's defense was that as stage manager, he did not organize the Putney show or print the programs.

Buffalo Bill's legal representatives were also on the lookout to serve papers against the Codys for representing themselves as Buffalo Bill Cody's son and daughter. Both S. F. Cody and Maud Lee had an understandable aversion to appearing in court after having faced terrifying proceedings in America—S. F. on the child assault charges, and Maud for the shooting of Mary King.

Fortunately, S. F. Cody (but perhaps not Maud) was able to hide in plain sight as an unbilled cast member of John Robinson Whitley's new international fair for the year 1891, the German Exhibition. Whitley's previous Earls Court shows had received lukewarm support from the British government; he was a private entrepreneur, and his shows were not publicly funded. The government was not greatly concerned over the political message that such shows delivered during the American and Italian Exhibitions, since relations with those countries were considered benign. However, Whitley's celebrations of France and Germany brought down open expressions of irritation from the royal family and from producers of British goods.

Whitley had developed a sincere belief that his national exhibitions were promoting peace and goodwill between Great Britain and other nations. Yet the fact that the arena shows in the Earls Court amphitheater focused on the thrilling martial heritage of each country seemed at odds with Whitley's faith in cultural exchanges as a tool of peace. The German Exhibition was the most notorious example: the arena show was nothing less than a tribute to German militarism through the ages. Edward, Prince of Wales, tried to discourage his nephew, Kaiser Wilhelm II, from offering any support to the German Exhibition. The kaiser, perhaps acting on Edward's request, offered only oblique encouragement.

When the German Exhibition opened at Earls Court in May 1891, the German representative at the inaugural ceremony was Prince Blücher von Wahlstatt, one of the imperial cousins less favored by the kaiser. John Whitley may have been an organizational genius, but he was an awkward diplomat. During the opening ceremony, Whitley introduced Blücher by making a lighthearted reference to the prince's great-grandfather, Marshall von Blücher, who had led the German army in alliance with the British army against Napoleon at Waterloo:

> The name of Blücher, in the stirring times of old, when allied with that of Wellington, was synonymous with victory, for together they engraved upon the tablets of history one sublime example of heroism—Waterloo.
>
> Once more that great historic name is heard, and once more a Blücher comes to the help of the British. You will all remember that, when the great Marshall first saw London, he gazed with longing eyes upon it, and exclaimed, "What a city to sack!"[18]

One can imagine the groans that must have traveled from Earls Court over to Buckingham Palace in response to Whitley's attempt at humor. Though the kaiser might have considered Blücher to be dull, Blücher graciously responded to Whitley by assuring the crowd that none of the old marshall's descendents had any intention of sacking London.

The exhibition's arena show, called "Germania," consisted of dramatic sketches from four periods in German history. The first scene was set in the eighth century. It depicted Teutonic tribesmen celebrating pagan rites, culminating in the sacrifice of a female Christian captive. The rites were interrupted by Saint Boniface, the apostle of the Germans.

He destroyed the pagan altar and, aided by his converts, freed the captive. Charlemagne's Frankish warriors arrived on the scene to vanquish the barbarians, and after a fierce battle, the pagans capitulated and converted to Christianity.

The second "life-picture" presented to the audience was a tournament from the Age of Chivalry in the fourteenth century. Before the entrance of the mounted knights, there was a maypole dance, quarterstaff bouts, wrestling on horseback, and comedic turns by jesters. In one sketch, the Franciscan monk Berthold Schwarz demonstrated his new invention—gunpowder—to one of the jesters. The armor-clad knights then faced off against one another, and the ultimate champion was decorated by Emperor Lewis.

After another break, the arena scene shifted to the Thirty Years' War in the seventeenth century. The hero of this act was General Albrecht von Wallenstein, and the action recalled was the famous Battle of the Alte Veste, in which the Swedish army of King Gustavus Adolphus tried to storm Wallenstein's camp at Nürnberg. Prior to the battle, Wallenstein's troops sang some lusty old German war songs while gypsy girls danced for them. Wallenstein fortified a derelict castle and dared Adolphus to assault it. The "Germania" enactment added a side story about a traitor spying on Wallenstein and his eventual capture and death—the same theme that had been included in the previous year's "Wild East" show.

The final act of "Germania" consisted of equestrian drills set to music, performed by the Imperial Guard, followed by a parade of men in the uniforms of contemporary German infantry and cavalry. Military parades and bands were especially popular in London in the 1890s, so this part of the arena show was aimed at the same audiences that attended exhibitions of military pomp. The foot soldiers and horsemen ended "Germania" with a hearty chorus of the anthem *Die Wacht am Rhein*.

S. F. Cody stated two years after the German Exhibition that he had performed as one of the Imperial Guard riders, but it is likely that with his horsemanship skills he also rode in the other sketches as well.[19] Maud never referred to the "Germania" show when recounting her career and, as will be shown, she likely was incapacitated during these months. If she did appear, it would have been as a maypole dancer or Hungarian gypsy dancer. "Germania" ran from May until mid-October 1891. The 1891 British census taken in the summer found the Codys living at a Hammersmith address, adjacent to the Earls Court complex.

Before "Germania" ended in October, evidence suggests that S. F. Cody and Maud Lee's marriage was strained. At some point during the previous months, the Codys met a woman named Elizabeth Mary King, nicknamed "Lela," along with two of her four children, sons Leon and Vivian. Mrs. King's father was John Blackburn Davis, a noted horse trader who sometimes provided mounts to the royal family. Her husband, Edward King, was a saloon owner. Edward and Lela had become estranged in the late 1880s.

Maud, apparently, did not immediately view Lela as a threat. At some point in the summer of 1891, she gave shooting lessons to young Vivian King. The fact that she willingly groomed the boy as a sharpshooter makes it seem plausible that the Codys were helping to further Lela's theatrical aspirations for her family.[20] Whether that was Lela's primary motivation or a mere pretext for a romantic involvement with S. F. Cody may never be known.

Just one week after closing his run in "Germania," on October 24, 1891, S. F. Cody was billed at the Winter Gardens Olympian Skating Rink in Birmingham. The advertisement read that Cody would exhibit pistol shooting and that there would be fancy shooting by "Master Vivian Cody (only eight years old), the most remarkable shot in the world." Two days later another revival of "Frank Hall's Burlesque of the Wild West" on skates started a two-week run at the same rink.[21] Later testimony by Maud Lee indicates that Lela King took over the role of Any O'Klay in the burlesque, and that Maud herself was not healthy enough to perform.

During the run of this show, on November 2, 1891, the court case in London that Buffalo Bill's representatives had brought against Frank Albert Hughes, the Putney show producer, was decided in Hughes's favor. Buffalo Bill's legal team failed to show up for the court appointment and therefore defaulted their case. If the Codys had worried that legal action against them would prevent them from continuing their careers in London, the news of this reprieve must have been welcome.

Yet the day after the Birmingham Wild West burlesque ended its run, Maud boarded the steamer *Scythia* bound for Boston.[22] Samuel F. Cody was not with her. When the ship's registrar entered Maud's name, occupation, and destination, she gave her name as Lillian Cody, heading for Pennsylvania. There was some confusion about her occupation: "Dressmaker" was crossed out and "Actress" written above it. Although Maud made it back to her parents' home in Swedeland, Pennsylvania, she arrived confused and disheveled, without luggage or money.[23] She would never see Samuel F. Cody again.

Frank Frayne, the first public performer of the mirror shot, accidentally kills his fiancée, Annie Von Behren, during a production of *Si Slocum*. (Engraving from the *National Police Gazette,* December 1882)

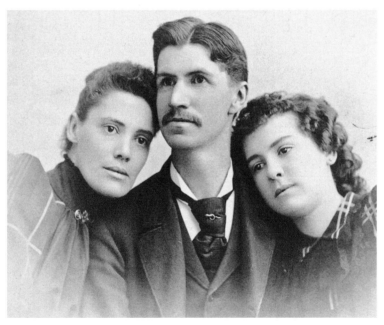

Jennie Fowler/Franklin/Moore/Kennedy/Coleman, aka "Mexis," pictured here with William B. Kennedy and daughter Josie, circa 1885. (McCaddon Collection of the Barnum and Bailey Circus, Manuscripts Division, Department of Rare Books and Special Collections, Princeton University Library)

Doc Carver (*left*) and Adam Bogardus (*right*) pose in Des Moines, Iowa, in early 1883 during their touring shooting competition. (Collection of the Trapshooting Hall of Fame)

Settlers in Oklahoma congregate at a land office to submit their claims. Harry Hill's sharpshooter performers, including S. F. Cody and his wife, Maud, were reported to be among those participating in the April 1889 land rush, although the Codys never filed a claim. (Photo LC-DIG-ggbain-02285, George Grantham Bain Collection, Prints and Photographs Division, Library of Congress)

Allen & Ginter cigarette cards of shooting champions (*clockwise*): Annie Oakley, Doc Carver, William F. Cody, and Adam Bogardus. (Allen & Ginter Album of World Champions, W. Duke & Sons Co. Collection, Rare Book, Manuscript, and Special Collections Library, Duke University)

Annie Oakley, although she did not originate the mirror shot, is now the shooter most associated with it. Here she reconstructs the position for a studio photograph. (Photo P.71.809, Vincent Mercaldo Collection, Buffalo Bill Historical Center, Cody, Wyoming)

Frank Butler, husband of Annie Oakley. Butler's presence at the deposition
of Maud Lee likely allowed him to piece together more of the true story of
her tribulations than any other person. He pitied Maud and blamed newspaper
reporters for the misidentification of her as Annie Oakley. (Photo P.71.113,
Vincent Mercaldo Collection, Buffalo Bill Historical Center, Cody, Wyoming)

One of the earliest existing images of Samuel Franklin Cody, published in the British entertainment magazine *Amusing Journal,* July 8, 1893. Note that he appears to be holding a mirror. (Courtesy of Jean Roberts)

HAMILTON BALL GROUNDS.

WEDNESDAY, JULY 10.

THE HISTORICAL OKLAHOMA

WILD WEST EXHIBITION,

Will give a performance Afternoon and Evening, Rain or Shine,

Traveling by its Own Train of Cars!

It is the greatest show of the kind ever placed on the road.

OKLAHOMA HARRY HILL

The Noted Scout and Boomer, is one of the best known, most desperate and daring scouts of the age.

CAPTAIN A. H. BOGARDUS,

Champion Wing Shot of the World, and his three Sons, and S. F. Cody, jr., and wife, Champion Revolver Shots of Oklahoma, are also with this show and appear daily.

20 Mexican Vanquoritas and Senoritas in fancy riding and roping.
50 Scouts and Cowboys.
A Herd of Buffalos and Wild Horses.

Grand Street Parade at 11 O'Clock

OKLAHOMA HILL.

An ad for Harry Hill's Oklahoma Historical Wild West Exhibition that appeared in the July 7, 1889, edition of the *Hamilton (Ohio) Daily Democrat.* Hill's show relied on remnants from Pawnee Bill Lillie's unsuccessful 1888 season. This rough drawing is one of the few known images of Hill.

CHAPTER 6

KING OF THE COWBOYS AND QUEEN OF THE LADY ROUGH RIDERS

1892–1895

Many years later, Maud's family put forward their own explanation for what happened to Maud during the five months between May and November 1891. They claimed that Maud had suffered injuries during a parachuting accident in England. The only known instance when the Codys performed with a parachutist was the Whit Monday program at the Half Moon Pub grounds in Putney, where the parachutist was Charles Baldwin. Newspaper accounts describing Baldwin's balloon ascension and descent, however, make no mention of other aeronauts nor any accidents.

At the same time, no mention of Maud can be found in any of S. F. Cody's engagements following the show in Putney with Baldwin the parachutist. For example, many years later when talking about her career, Maud listed the Wild East show of the French Exhibition, but not the German Exhibition.[1] So it does appear likely that she was incapacitated between May and November, but whether that inactivity was caused by a physical injury, as her family claimed, is questionable.

Years later, Maud's family declared that S. F. Cody had abandoned Maud. They said he promised to follow her back to America by Christmas, and that he put a sick woman on a ship home with no money and only the clothes on her back. Yet they never pursued any legal action or claims against Cody until after he died, even though doing so would have been a reasonable course of action had he abandoned Maud after an accident, as they claimed.[2] Moreover, long before Cody passed away, news of his later career fame must have come to their attention.

Clues exist to another possible explanation of what happened to Maud and her marriage. One factor may be that Maud experienced a worsening case of mental illness, leading S. F. Cody to decide that he could no longer care for her. When Maud arrived back home in Swedeland in November 1891, her family described her as "very worried," saying that her health declined at a fast pace and that she became "alarmingly ill"; but they made no mention of doctors, a diagnosis, or hospital care. They may have recognized Maud's symptoms immediately, because Maud's incapacity parallels similar health problems in her mother.

Little is known about Maud's parents, Joseph and Phoebe Lee, but what is known is curious indeed. Joseph Lee, born about 1852, was a common laborer who hailed from Bollington, Cheshire, south of the industrial city of Manchester, England. He married Phoebe Nadin, born about 1851 in Ashton-under-Lyne—just east of Manchester. Phoebe had a younger sister, Elizabeth Lillian Nadin, born about 1858. Phoebe and Elizabeth Nadin also came from a humble background, and both worked as card-room sorters in a cotton mill near Manchester. Elizabeth later married a coal miner named William Kay, but he died young.[3]

Joseph Lee and Phoebe Nadin married before the age of twenty and immigrated to Canada. At Innisfil, Ontario, their daughter Maud Maria Lee was born on February 22, 1872. Joseph's occupation when Maud was born was listed as common laborer.[4] The Lees, despite their subsistence income while at Innisfil, decided to go back to England at some point in the 1870s. They reappeared in the 1881 British census as residents of Aston-under-Lyne, Phoebe's hometown. Three years later, in 1884, Joseph immigrated to Pennsylvania by himself. The following year, in 1885, Phoebe and Maud joined Joseph in America.[5]

Among the possible reasons for the moves back and forth across the Atlantic, the most plausible one is that Phoebe Lee was sick and unable to care for herself or her daughter. Once Phoebe's health improved somewhat, and Maud was old enough to help look after her mother, the family was reunited in America. After Joseph, Phoebe, and Maud settled permanently in America, Phoebe's widowed sister Elizabeth came over from England to live with them near Norristown. It may be that Elizabeth did so to help care for Phoebe and her daughter.

Phoebe Lee died in 1910; a short while afterwards, Joseph married her sister Elizabeth. Therefore Maud's aunt became her stepmother. Phoebe Lee's funeral notice in the *Philadelphia Inquirer* invited mourners to visit the home of "her brother," Joseph Lee. Either this was a bizarre reporting error or Phoebe had long been so ill that she was introduced to others as Joseph's sister, and Elizabeth and Joseph had posed as husband and wife for many years. A few months after Phoebe's death, they relocated to Camden, New Jersey. They continued to live together as husband and wife for twenty more years after Phoebe's death.[6]

The disease known today as schizophrenia most often first manifests itself in women between the ages of twenty and thirty, the age range when both Phoebe and Maud became ill. There is not always a family history of schizophrenia—that is, it is not purely a genetic disorder—but schizophrenics are ten times more likely to have a schizophrenic parent than non-schizophrenics are.

Physical factors, such as stress, poor nutrition, and drug abuse, are known to trigger episodes of the disease. During their lives, neither Phoebe nor Maud Lee was diagnosed with schizophrenia—the term was not yet clearly defined nor in general use. Even today, there is not universal agreement about whether schizophrenia is a single condition or an umbrella term used to describe a set of behaviors. The medical term used at the time when Maud was finally diagnosed was even more generic: "dementia praecox."[7]

An incipient mental illness and a possible physical injury were not the only ailments that might have afflicted Maud Lee in the summer of 1891. She was, by her own later admission, addicted to the opiate drug morphine.[8] One or more of these conditions may well have been linked. For instance, Maud could have been administered morphine for pain following a physical injury, or the stress of a physical injury and longtime drug use could have brought on an episode of schizophrenia, or Maud might have been given morphine to calm the agitated mental state of a schizophrenic episode. S. F. Cody, a teetotaler during his career, probably would not have allowed Maud to use firearms while under the influence of morphine.

Morphine was not illegal in the 1890s, in either the United States or Great Britain. It had been isolated as a crystalline salt from opium by a German pharmacist in 1805, but did not enter into widespread use as an analgesic until the development of the hypodermic needle in the 1850s. Morphine was also an ingredient in numerous nonprescription patent medicines widely marketed both in America and in Britain. Because patent medicines disguised their ingredients, however, few users would have been aware that the "effective additive" in their drink was morphine. Still, most addicts became hooked through intravenous injections of morphine obtained from doctors or other addicts.

Morphine addiction had reached alarming proportions by the 1890s, and Maud Lee fit many aspects of the typical user profile. She was a white, Anglo-Saxon woman between the ages of eighteen and forty. Many women addicts were middle-class homemakers, but Maud fit another significant demographic: she was part of the theatrical community. Widespread morphine use among the theatrical and literary communities was common knowledge. Accessories for carrying vials and hypodermic syringes were sold in fashionable stores; and it would not have been unusual to see injections being taken in open view in certain saloons.

Any number of people could have recommended morphine to Maud as a cure for the pain of a physical injury, for the mental agitation and dissociation of a schizophrenic episode, or even for the mere stress of touring with a traveling show. Prolonged drug abuse is a known trigger of schizophrenia, but the short-term effects of morphine have the opposite effect of appearing to calm the outward agitation of a schizophrenic episode. In other words, Maud might have been given morphine to alleviate her symptoms, which in fact, could have done irreparable harm to the more serious underlying condition.

In the early months of 1892, when Maud was trying to recover from the ailments that forced her home to her parents, a neighbor of the Lees came to their house carrying a newspaper from London—it was probably the *London Era.* In it was an advertisement for a variety program at the

Royal Aquarium, London. Listed among the acts was "S. F. Cody and family, Rifle Experts."[9] Samuel's new family consisted of Lela; her three sons Edward King, Jr., Leon, and Vivian; and her daughter, Liese (who, unlike her brothers was not a performer). Maud Lee feigned shock at hearing this news; perhaps she was genuinely surprised at the speed with which Lela had usurped her. She could not have been totally unaware of the situation, however, for before leaving England, she had an open confrontation with Lela.

According to statements Maud made many years later, Lela had told Maud that the latter was herself to blame for losing the Any O'Klay part in the stage burlesque because she was addled by drug use. Furthermore, Lela informed Maud that she intended to divorce Edward King in order to marry S. F. Cody, and that she, unlike Maud, would be a good, Christian wife to Cody.[10] Maud, still just nineteen years old and ill, did not have the strength to face down her older rival.

Whether or not Cody was present during this scene is not known, but he must have arranged for Maud's passage back to America. He was probably well aware that Maud's problems went deeper than morphine addiction. Maud's mental illness and drug addiction would explain why her family never made a public effort to label Cody a bigamist while he was alive: they knew that Maud's problems were a contributing factor.

One of the most terrifying facts about schizophrenia is that it is a progressive disease. After the first onset episodes, schizophrenics can recover some normal function. However, they are aware that they are different, and live in fear of slipping into the next episode and of losing loved ones. Morphine addicts, too, can recover from a physical addiction after going through a painful, harrowing withdrawal, but are psychologically prone to fall into use again. Maud Lee faced daunting challenges to regaining a normal life.

By 1892 she had already lost Cody, but had otherwise defined herself as a Wild West show performer—an actress, equestrienne, and sharpshooter. Her marriage was gone, but she still had a career to cling to. She had chosen a difficult profession that required grueling travel, physical risk, excellent hand-eye coordination, attention to detail, balance, strength, and fearlessness. During stunts where assistants held targets, those people would place their lives in Maud's hands. A sharpshooter is the last profession in which one would hope to find someone with a deteriorating mental condition, but the episodic nature of onset schizophrenia is insidious. Nor had Maud totally kicked her morphine habit, which also has a high rate of recidivism. Target shooting and trick riding were risky ventures for a fit person; in deciding to continue along the same career path given her physical and mental state, Maud Lee seemed headed for disaster.

After Maud returned to America alone, Samuel F. Cody and his new soul mate, Lela King, wasted no time retooling the sharpshooting act. Lela replaced Maud as target holder and occasional rifle shot. Her sons Leon and Vivian, who had first been introduced to shooting by Maud, also displayed their marksmanship in the act.

In bringing the boys into the act, Cody was taking a page from his Oklahoma Wild West mentors, Adam Bogardus and Harry Hill, both of whom incorporated their sons into their demonstrations. Cody and Lela had posters printed to advertise "S. F. Cody and Family, the Champion Shooters of America." Their first performance together was at a roller-skating rink in Leeds, England, during the week of November 26, 1891—less than three weeks after Maud had left for America.[11]

It is not known why Lela's husband, tavern owner Edward King, allowed his family to be hijacked by Cody. In later published interviews, Cody fabricated an elaborate tale of how he had met Lela before 1891, while traveling back and forth between the United States and Britain to train horses for her father; and how they had married and moved to Texas, where Leon and Vivian were born. There is however no proof that Cody ever traveled to England before June 1890. If in truth they met as a result of the equine trade, then it is more likely they did so when Cody was handling horses during the French or German exhibitions, as steeds had to be procured, replaced, and released in each of the two years. Since Lela's father died in 1890, she may not even have been involved in his business.[12]

Any acquaintances of the King or Davis families could have discredited Cody's story, but instead they kept silent. Census records of the 1880s showing Lela's children living in different households indicate that Lela and Edward King had separated some time before she met Cody. If Lela had been victimized somehow by Edward King, he was in no position to expose her history without bringing unwanted attention to his own behavior.

After a few more engagements in provincial venues, Cody, Lela, and the boys brought their act back to London. They began an engagement at the Royal Aquarium in March, and played there nearly continuously through to the Bank Holiday in August. The Royal Aquarium had been built with higher ambitions than serving as a music hall. Opening in 1876 as a place for highbrow intellectual entertainment and instruction, it featured a cast-iron and glass-barrel roof (the Crystal Palace influence) supported by walls of Portland stone, with an ornate columned façade and cupolas over the entrances.

The glass roof allowed palm trees to survive inside the main hall, which also contained a sculpture garden, huge saltwater aquariums, an orchestra pit, an art gallery, reading rooms, dining rooms, smoking rooms, a roller-skating rink, and a small theater. The facility was mismanaged almost from its inception, and proved a financial disaster. Over the years, the owners brought in more variety acts, and therefore by the 1890s most of the water tanks, palm trees, sculpture gardens, and such were gone, and the Royal Aquarium was known primarily as a music hall. The largest glass tank was preserved, not for exotic fish, but for an attraction featuring pretty young women swimming underwater.

During the same months that Cody and the Kings were appearing at the Royal Aquarium, Buffalo Bill's Wild West came to London for the first time since 1887. Though audiences were now familiar with the show and its stars, William F. Cody's and Annie Oakley's return proved as popular as ever. Buffalo Bill's solicitors refrained from further harassment of S. F. Cody, but Oakley's husband, Frank Butler, was vigilant against imitators. Butler wrote back to America that England was "overflowing" with sharpshooter "humbugs." He accused these acts of using blank cartridges and trickery, dismissing them outright: "There are a great many so-called champions tolerated in England that would not last long in America."[13]

In late 1892, Cody and his new family crossed the English Channel and were booked to appear in one of the most famous Parisian music halls of La Belle Époque, the Casino de Paris. They appeared during the height of the popularity of French music halls, at a time when those places were being celebrated and immortalized by masters of modern painting: Toulouse-Lautrec, Degas, Renoir, and Manet, as well as Picasso and Cassatt. Parisian audiences embraced the Codys and, fresh from their success in Paris, they returned to London with enhanced fame. After another run at the Royal Aquarium in July and August 1893, they again voyaged to France in late summer for a more prolonged tour of other French cities.[14]

While in Nice, S. F. Cody indulged in one of his small vices, the gambling table. As a marksman, Cody would have been well versed in the art of making bets, but should have been wiser than to approach casino games thinking he knew of a system that would beat the house odds. He lost heavily at Nice after putting his faith in such a system, most likely a variant of the Martingale system, at the roulette table.[15]

The Martingale system advocates doubling the stakes after a lost bet, since the odds are very high against a long string of losses on a basic binary bet—heads or tails, red or black. This system has two fallacies: first, roulette is not binary—the green 0 and 00 give an advantage to the house; and second, the odds against a string of defeats are not as high as many people perceive, making it easy to expend all one's chips before reaching the needed winning bet. Many bettors also mistakenly believe that a string of losses improves their odds on the subsequent bet, which is a logical fallacy.

If Cody had applied a Martingale system to his own sharpshooting, he would soon have lost his nerve shooting at targets held by Maud, Lela, or the boys, since the system would predict that the risk of missing his aim rose with each performance. Instead, his daredevil instinct subliminally informed him that the odds against mishap were the same for every show, given his skill and the reliability of his firearms. That he could ignore this lesson in probability indicates the seduction of the casino atmosphere.

His casino losses notwithstanding, S. F. Cody's wagering pastime presented another opportunity that proved much more lucrative: races pitting horse against bicycle. Side bets in private exhibitions were part of a western sharpshooter's income. During the 1880s, Buffalo Bill Cody did not forbid his employees from engaging in these exhibitions, creating opportunities that even Annie Oakley did not pass up. During the first visit of Buffalo Bill's Wild West to Great Britain in 1887, Cody encouraged two of his cowboys, Broncho Charlie Miller and Marve Beardsley, to participate in a six-day endurance horse-versus-bicycle race that took place inside the Royal Agricultural Hall at Islington.

The idea of racing bicycles against horses is virtually as old as the bicycle itself, if you accept the addition of foot pedals to the hobbyhorse as the birth point of the bicycle. In July 1868, a bicycle built by A. Favre raced against a horse and buggy along a forty-five-mile stretch of road ending at Toulouse, France. Seven years later, British cyclist David Stanton raced his high-wheeled ordinary against a horse and buggy along a ten-mile track at London's Alexandra Palace, a popular amusement center. Stanton lost by just forty seconds. A few months later, in the fall of 1875, a seven-hundred-mile Paris-to-Vienna race pitted a high-wheeled bicycle against a horse, and the bicycle triumphed by a margin of days. At a track in Toronto, Ontario, a popular bicycle racer named Ernestine Bernard won a three-mile contest against a racehorse. Male spectators might have been attracted by Ernestine's form-fitting cycling costume—an early example of how the popularization of the bicycle helped to break conventions in women's dress, culminating in the athletic bloomers of the 1890s.

By 1887, both the American and European public had been repeatedly exposed to horse against bicycle races. Buffalo Bill's troupe added the extra interest of pitting expert western riders—both Charlie Miller and Marve Beardsley were veterans of the legendary Pony Express—against accomplished high-wheeled racers. High-wheeled bicycles, or ordinaries, were favored by racers and sportsmen as the fastest design, since power was applied directly to a large-circumference wheel. They were difficult to mount, however, and had poor weight distribution, often causing headfirst tumbles.

Those not daring enough to risk riding ordinaries—women, the elderly, and members of the upper class who felt the need to maintain decorum—chose to ride tricycle designs. It was on tricycles that the first chain drives were featured. The chain drive quickly led some inventors to rethink the design of the bicycle: by gearing up the chain from the pedal to the wheel, it was possible to make a small wheel perform like a much larger wheel. The breakthrough design pinpointed by most bicycle historians is the Rover bicycle, first marketed in 1885.

The Rover was a low-mounting, small-wheeled bicycle with the chain drive relocated to the rear wheel. Compared to an ordinary, it had better weight distribution and less wind resistance. On the other hand, it was heavier than a racing ordinary, had a rougher ride, and was more expensive. For these reasons, racers were not at first interested in the Rover safety bike design, although it became very popular with the general public.

Two critical improvements occurred just after 1887 that spelled doom for the high-wheeled ordinaries: in 1888, John Dunlop introduced pneumatic tires, which increased the speed of safety bicycles by about a third compared to solid tires; and in 1890, the introduction of the strong, diamond-shaped frame made the ride much smoother. In 1890, bicycle racers finally started to adopt the safety bicycle. Just three years later, ordinaries were considered relics in the world of bike racing. Safety bikes took America and Europe by storm, and bike sales doubled from 1887 to 1890, and then doubled again in just one year, 1891. Cycling became a popular craze. During the 1890s, the American recreational journal *Sporting Life* had two major sections: baseball and bicycling.

In the midst of this 1890s fad, S. F. Cody was reminded of the 1887 race made by Buffalo Bill's men—or perhaps he had seen the roller-skater-versus-bicyclist races at Olympia in 1890. At any rate he was astute enough to recognize an opportunity to use his horsemanship to entertain a crowd, and perhaps to place side bets while he was at it. In October 1893, S. F. Cody arranged to race against a noted cyclist, C. Meyer of Denmark, at the Paris trotting club. In the posters publicizing the race, Cody pronounced himself the "King of the Cowboys."

The twelve-hour race took place on Oct. 29, 30, and 31, 1893, four hours each day. Cody was allowed to use four different mounts. Cody won Friday's trial; Meyer won on Saturday; and Cody won again on Sunday, giving him a grand total of 217 miles to Meyer's 206 miles. Cody's triumph was met with cheers, and he took a victory lap around the course standing astride two horses, a trick he lifted from Adam Forepaugh, Jr.'s, climactic act in the 1888 circus shows.[16]

Following this success, Cody arranged another race in Paris on November 12, 1893, against tandem cyclists Henry Fournier and H. Gaby. This time Cody was allowed six different mounts, and the race was divided into two days of two hours the first day and four hours the second. Because this race took place at a velodrome, the cyclists bene-fited from its much smoother, banked track. Cody, on the other hand, had difficulty negotiating the sharp turns with his ornery mounts, and was even thrown to the ground once. The grand total of the two days found the tandem duo the victors, 178 kilometers to 173 for Cody.[17]

For his next race on December 23, 1893, Cody moved the venue to Bordeaux and found sponsors for a cash prize of 2,500 francs. His oppo-nent was the famous cyclist Henri Loste. The format was once again twelve hours of racing over three days, four hours each day. Presumably, Cody took more attention in selecting the mounts and the track condi-tions, and negotiated the use of ten different mounts. With these variables in his favor, Cody emerged the victor.[18]

The Bordeaux race cemented Cody's fame throughout Europe. He, Lela, and the boys put together a new tour of France in the spring of 1894 that featured horse-against-bicycle races. Since he negotiated with different opponents at each stop on the tour, Cody had the advantage of knowing the factors that would tilt the odds in his favor. By this time, he also traveled with his own stable of horses, which he had specially trained to allow for fast mounting and dismounting.

In August 1894, Cody's tour took him to Munich, Germany, to meet the famous cyclist Josef Fischer in a three-day series of races. In addition to the popular hype Cody received for moving his challenge to Germany, a great deal of interest in this race was shown by German army officials, who were studying the possible military applications of the safety bicy-cle. For roughly twenty-five years—the period between the advent of the safety bicycle in the early 1890s and the end of World War I—modern military powers experimented with the use of bicycles in war. Compared to horses, bicycles were lightweight and compact, needed no food, were easily and quickly mended, and needed no housing. Even after motor-ized vehicles became reliable, bicycles had the advantage of not need-ing fuel or wide paved roads, and of running silently. Some armies tried folding models that could be carried by infantrymen or (after World War I) paratroopers. Many armies were quick to introduce bicycles, at least for scouting and messenger duties, if not in fighting units.

When alerted to the German military's interest in bicycles, S. F. Cody no doubt filed away the fact that even technology developed for sport and entertainment might interest generously funded military engineers. It was a lesson that would play a large factor in the rest of his career. The cyclist Fischer beat Cody decisively in the Munich match, which likely only impressed Cody further that the world was changing, and new frontiers were opening in mechanics' workshops, not on the wild prairies.[19]

Although the races generated international fame for Cody and proved lucrative, it was not long before the sponsors of legitimate bicycle races—national cycling unions—realized that these stunts could be easily manipulated. If a gambling scandal erupted that involved any of Europe's star cyclists, the whole sport would be tainted for years. These organizations began to dole out penalties to any cyclists who agreed to compete with Cody. The term they applied to these events was "hippodroming"; that is, participating in contests with predetermined winners.

For example the British magazine *Cycling,* in the January 25, 1895, edition, noted that the triplet team of Coquelle, Lamberjack, and Tricot, who had recently raced against Cody in Brussels, had been disqualified from racing by the Belgian and French Union for six months.[20] Mentions of Cody's stunt races in the pages of *Cycling* began with a tone of bemusement in 1894; by 1897 he was treated with hostility and not a little chagrin. On the one hand, Cody debased the legitimacy of impartial competition, but on the other hand, he exposed thousands to recreational cycling.

It is not known how long Maud Lee took to recover from the broken marriage, agitated mental state, and morphine addiction she exhibited on her return from England in late 1891. Most schizophrenics recover after their first onset but have another episode within five to seven years. In whatever manner Maud's illness was recognized by her family, one reassuring fact was simply that she returned to her family to recuperate. Attentive family care is now recognized as a significant factor in the treatment of first-onset schizophrenia.

Only a few vague clues exist to Maud's lost months after her return from England. She later mentioned that during her career she had danced in variety halls in winter months and even led a troupe of dancers. She also stated that she had performed in another minor Wild West touring show, the name of which cannot be confirmed. More ominously, she also admitted to having been in and out of sanitariums to seek treatment for her drug habit. It is not known whether any of these events took place in 1892 and 1893.[21]

By the summer of 1894, Maud was in Iowa, performing a shooting and riding act at local and state fairs. It seems more than a coincidence that Maud's public life resumed in Iowa, the home state of S. F. Cody's mother, Phoebe Cowdery. Perhaps Maud returned to visit Mrs. Cowdery, with whom she and S. F. had stayed in late 1889 and early 1890. Maud knew that Phoebe Cowdery had been abandoned by her own husband, S. F. Cody Sr., in the 1870s, so perhaps Maud prevailed upon Phoebe to contact her son and implore him to return to his legal wife.

The pre-event publicity for one of the fairs in which Maud appeared claimed that Lillian Cody (Maud's stage name) had been with Buffalo Bill's Wild West at the Chicago Columbian Exposition of 1893. This was just one example of show agents exaggerating their performers' backgrounds. On another occasion, a show agent used posters depicting Annie Oakley with the name "Lillian Cody" pasted below the image. The fact that various managers and agents knowingly acted to conflate her stage identity with Annie Oakley's would play a major role in Maud's future.

Agricultural fairs offered another venue for Wild West performers when salaried jobs in touring shows and vaudeville circuits could not be found. Annie Oakley herself made several appearances at fairs—which also provided opportunities for shooters to make money on side bets by facing local marksmen. Lillian Cody announced a challenge of her own: for $100, she would ride the wildest horse brought before her. This was a boast that even a trainer born in the saddle would hesitate to make, but evidently Maud had been a quick student of S. F. Cody's equine-handling skills.

In August 1894, Maud appeared at the three-day Shenandoah Athletic Tournament in Shenandoah, Iowa, along with footraces, tugs-of-war, baseball games, tennis matches, and bicycle races.[22] At the end of the month, she made her way to Des Moines for the biggest billing of her solo career: the Iowa State Fair.

Agricultural fairs in the United States developed from the tradition of Old World harvest fairs but introduced an educational element intended to teach landowners new scientific methods of production. Many new settlers had only a rudimentary understanding of crop and livestock management, and the American soil, weather, and other aspects of the native ecology challenged their European-based skills.

By the mid-1800s, statewide fairs were being held in northeastern states. Competitive exhibits to show off produce and farm animals also began to encompass handmade arts and crafts. The entertainments offered by these agricultural fairs began modestly, usually with equestrian events. Horse races were popular but controversial, since they inevitably attracted gambling—a mixed blessing to promoters also trying to attract morally sensitive fairgoers. Female equestrian events were an early drawing card in state fairs, especially to male crowds who may have been judging the women's forms as much as their horsemanship.

Dating from the 1850s, the Iowa State Fair grew over the decades to become one of the largest fairs in the country, and eventually became the quintessential American state fair, immortalized in Phil Strong's 1931 novel *State Fair,* in which the temptations of the fair's side entertainments constituted a dark theme. Later, in 1945, the Rodgers and Hammerstein musical of the same name portrayed a more romantic interpretation of Strong's book. In 1894, however, the Iowa State Fair was struggling to bounce back from a disastrous 1893 season. Attendance in 1893 had plunged due to a national economic depression and the loss of patrons to the extravagant Columbian Exposition in Chicago, just one state to the east. To combat the drop in attendance, the Iowa State Fair producers placed increased emphasis on amusements: they licensed dozens of food caterers, sideshows, balloon ascensions, daredevil acts, horse races, and other grandstand shows—including the riding and shooting of Lillian Cody.

Maud was billed as the "Queen of the Lady Rough Riders," the term "rough rider" meaning a rider of unbroken horses.[23] She performed her riding and shooting act at the fair's amphitheater, in front of large, appreciative audiences. Whether she made good on her bet to ride any wild horse brought before her is not known, but the Iowa State Fair's annual report does mention her trick shots: shooting a lemon off her assistant's head, splitting a playing card held edgeways, shooting glass balls thrown in the air with a pistol, and performing the mirror shot.[24]

At the same time—August 1894—her estranged husband S. F. Cody, hailed as the "King of the Cowboys," was racing against cyclist Josef Fischer. For a brief moment, both S. F. Cody and Maud Lee achieved their once-joint ambitions to become Wild West stars. The applause of the crowds, in that summer of 1894, must have made their hearts swell. Cody was the "King" and Maud, the "Queen," but they were oblivious to each other's achievements. They were separated by thousands of miles and the even larger gulf of a failed marriage.

For S. F. Cody, the success of his shooting act and horse-racing events drew him closer to Lela and her children, his new family, and allowed him the freedom to realize other long-held ambitions. Maud's brief taste of show business fame did not last. In September 1894, Maud performed her act at the Michigan State Fair and was paid a sum of $100.[25] Later in the fall of 1894, she was scheduled to appear at the Texas State Fair, which rivaled Iowa as a top annual agricultural fair. As the Texas State Fair opening in mid-October neared, however, no confirmations of Lillian Cody's appearance were found. Instead, the pre-fair publicity announced an exhibition of shooting by Doc Carver, the "Evil Spirit of the Plains."[26]

After the Iowa State Fair, Maud M. Lee, aka Lillian Cody, disappeared again for another two years, her name missing from any public records, news accounts, and stage billings. Perhaps an explanation of her whereabouts during that time will come to light; but lacking that evidence, the traces of her public life are consistent with the episodic pathology of her afflictions. Medical science, in 1894, offered no cure for schizophrenia or for the recidivism of opiate use. Nevertheless, Maud Lee's performing career was not quite finished.

Chapter 7

Arena to Stage

1896–1900

From Germany, Samuel F. Cody, Lela, and her children continued their tour of Europe for another two and a half years, from 1895 through to mid-1897. Their engagements combined shooting exhibitions with horse-versus-bicycle races. In early 1895, their travels took them to Belgium, where they became popular fixtures of the Circus Lenka in Ghent.[1]

The Codys spent the remainder of that year in Germany, Italy, and Switzerland. It was during this period that S. F. Cody acquired a new tool for use in his act: a radical handgun designed by Hugo Borchardt.[2] The Borchardt C-93 model was a semiautomatic pistol that employed a toggle-lock cartridge-loading mechanism patented by machine-gun inventor Hiram Maxim. Cody was impressed by the C-93's rapid-fire capabilities. Today, it might be called a machine or assault pistol.

The Borchardt C-93 design inspired a more famous gun, the German Luger P08 pistol, which was first produced a few years later. To the modern eye, images of S. F. Cody shooting the Borchardt seem jarring: they show a man in the garb of an American frontiersman holding a weapon that many people associate with World War I or World War II Germany. The Borchardt was a fine target weapon, but its bulk and heavy recoil prevented it from being mass-produced.

It is tempting to view Cody's fascination with the Borchardt as a symptom of the waning of his American character, but this would be wrong on several counts. First, settlers of the American frontier embraced new technologies as quickly as people anywhere. Second, Cody proudly held fast to his other Americanisms: his profane speech, his manner of dress, and his sense of humor were unmistakably Yankee. Third, the Borchardt gun had a strong American heritage of its own. Sir Hiram Maxim, the inventor of the automatic toggle-loading mechanism, was born in Maine in 1840 and lived in America until 1881. Hugo Borchardt spent most of his adult years in the United States, working for sewing-machine and firearm manufacturers in Massachusetts and Connecticut.

It was also in 1895 that S. F. Cody fathered his only natural child, Franklin Leslie Cody. Young Franklin was born on September 7 in Basel, Switzerland. Lela seems to have demonstrated an odd indifference towards young Franklin, as evidenced by the fact that he was left with nannies in Switzerland during his infant years while the rest of the family, including Lela, continued to tour. She was rarely photographed with the child, in contrast to the several posed studio portraits made of father and son.

There is however no evidence that Lela was not Franklin's natural mother. S. F. Cody was known to be flirtatious with other women, but no hard proof exists that he was unfaithful to Lela, his common-law wife. In several instances, over all their years together, she demonstrated absolute trust in Cody.[3] He was most certainly a devoted father figure to Lela's children. He came to depend upon them in all his endeavors, and they responded with adoration, calling him Dad.

The Cody family spent all of 1896 touring Italy and the Mediterranean: Catania in Sicily, Sliema in Malta, and Tunis in Tunisia. In addition to shooting demonstrations and horse races, the Codys added other spectacles to their shows, such as chariot races and wrestling bouts. S. F. Cody did not want to lose his association with the American frontier, however, and to that end started to develop dramatic western scenes and sketches. Many of these were familiar elements of Wild West shows, with themes of villainous traitors, endangered women and children, savage Indians, and a heroic white scout. Cody was plainly working toward a longtime goal that dated back to his days with Maud: starring on stage in a full-length western melodrama.[4]

With this goal in mind, Cody, Lela, and the family began a slow return to England through France in 1897. Cody billed it as a "farewell" tour, and on occasion invited some of the cyclists against whom he had raced in 1893–94 to a nostalgic rematch. The Codys retrieved Franklin, now a toddler, from his nannies in Basel, Switzerland, and returned to England that summer. In August they sponsored a series of popular bicycle races at Wood Green and Kensal Rise in London, most of which involved amateur cyclists racing against other amateur cyclists, but did include at least one race where one of the Codys would race on horseback against cyclists.[5]

By 1898 the Codys had settled at the Alexandra Palace recreation center at Wood Green in North London. The Alexandra Palace was another grandiose Victorian structure modeled after the Crystal Palace and enclosing seven acres under its roof. There were over 160 more acres of parkland surrounding the building, on which were built sporting fields, gardens, ornamental walks, lakes, ponds, and other recreational facilities. The general manager of the palace, Cecil Barth, was encouraged by the crowds that Cody had drawn for his large bicycle races of the previous year; and he agreed to set up an outdoor arena in which the Codys could perform their act.

Descriptions of this show make it clear that the Codys were trying to refine some of their dramatic sketches.[6] Cody had likely spent the winter of 1897–98 writing a five-act stage melodrama. As a result, by the summer of 1898 he was preparing stage scenery and props in an area at Alexandra Palace set aside for him by general manager Barth. At the same time, Cody could not resist tweaking his shooting act, adding a stunt that brought him into contact with another performer who had become a fixture at Alexandra Palace, the balloonist Auguste Gaudron.

Gaudron, a Frenchman, was a kindred spirit to Cody. He, too, had been born outside of Britain, and had arrived in 1890. Like Cody, Gaudron had started his performing career in a touring circus before striking out on his own. While Cody tempted fate by shooting at loved ones and being shot at by them, Gaudron risked death as a balloonist and a parachutist—endeavors that at the time carried a high mortality rate. Gaudron had several close calls that he lived to tell about,[7] but some of the parachutists he employed were not as lucky. If the two men ever confessed to each other the events in their lives that haunted them, Gaudron could match Cody's story of Maud Lee's woes with his own tale of Louisa Maud Evans.

In July 1896, Gaudron had placed an ad in the *London Era* to recruit a new aeronaut. He was approached while on tour in Cardiff, Wales, by a young woman who gave her name as Grace Parry. She told Gaudron that she was more than twenty years old and had experience as a gymnast. She also said that she had made an ascent in a tethered balloon before, and had a cousin who had made parachute jumps. She was eager to try, and in fact said that she had been hoping to make a jump for the past four years.

They agreed that she would attempt her first jump on the evening of July 21. Even at 8:00 P.M., there was still plenty of light in the sky. Gaudron hooked her suit to the parachute cords and checked the equipment. The balloon had been inflated, and the young woman sat in a sling that dangled underneath it. The canopy of the parachute was fixed to the side of the balloon by a release snap. Gaudron spoke to her and saw no evidence of hesitation. He put a cork life preserver around her waist as a precaution, although the wind was blowing over the land. He told her to jump after passing over the Cardiff infirmary.

The balloon was let go and she ascended about two thousand feet, then jumped. She descended slowly, but a crosswind caused her to drift towards the sea. No one saw her moving her arms or legs, but to Gaudron's eye she appeared conscious as she came down into the Bristol Channel. Volunteers manned lifeboats to recover her, but she was not found before dark. Her body was later found on the sands near the East Usk Lighthouse.[8]

At the inquest, it was revealed that shortly before ascending, she had told a young man that she was troubled and did not care whether she came down from the sky or not. It was also revealed that her name was not Grace Parry but Louisa Maud Evans, and that she was an orphan who was employed as a companion to the wife of a circus owner. Moreover, she was only fourteen years old.

The jury determined that her cause of death was accidental drowning. They also ruled, however, that Gaudron had demonstrated great carelessness and lack of judgment in allowing her to make the ascent. When their verdict was read, Gaudron burst into tears, turned to face the wall against which he was standing, and sobbed until led out of the room.

From that incident forward, Gaudron aspired to be something more than a mere daredevil showman. He and a new young partner, Harry Spencer, developed a keen interest in the potential of aeronautical engineering. They had a workshop at Alexandra Palace and were involved in some innovative experiments during the same period that the Cody family was performing at Alexandra Palace.[9]

S. F. Cody came up with a new idea to impress his audience—shooting model hot-air balloons after they had ascended several hundred feet. Helium-filled latex balloons were not yet available, so Cody undoubtedly sought out the assistance of Spencer and Gaudron to craft these balloons. Gaudron's print advertisements claimed small-scale balloons as one of his specialties. The extent of conversations among Gaudron, Spencer, and Cody may never be known, but given Cody's later endeavors, the proximity between these fixtures of Alexandra Palace in the summer of 1898 suggests that they introduced Cody to the realm of aerial navigation.

In 1896, less than two years after her appearance at the Iowa State Fair, Maud Lee received top billing in a new Wild West attraction in Cincinnati, Ohio. This show, called "Historical Cincinnati," was commissioned by the Cincinnati Zoological Gardens to illustrate western life.[10] In fact, the show was designed from the onset to be a showcase more for eighty-nine members of the Sicangu Lakota tribe than for any white performers. The Sicangu performed the same standard skits seen in Buffalo Bill's Wild West—the Battle of Little Bighorn, the attack on a stagecoach, and so on—but they also presented their own reenactment of the massacre at Wounded Knee and a nighttime Ghost Dance. The Sioux also welcomed the public to walk through their encampment during the daytime.

A year earlier, in 1895, a group of Cree Indians attached to a Wild West show had been stranded across the river in Kentucky when that show collapsed. They were invited to camp on the zoo grounds until their passage back to Montana could be arranged. While the Indians remained there, the zoo experienced an increase in visitors, who came to gawk at the Native camp. The Cree camp soon brought in enough profit to help defray the cost of their trip home. This impressed the Zoological Society with the idea of inviting another American Indian encampment in 1896.

Because of experiences like those of the abandoned Cree in 1895, however, the federal Office of Indian Affairs was reluctant to allow groups of Indians to leave their reservations to tour with Wild West shows. The government's ulterior motive was to discourage Indians from clinging to tribal traditions, and to pressure them to assimilate into white society. To placate government concerns, the Zoological Society agreed to guarantee the safe return of the Sicangu Sioux to their Rosebud Reservation in South Dakota by means of a $10,000 bond.

The zoo's director, John Goetz, justified the public display of the Sioux by claiming that ethnological exhibitions were within the mission of the Zoological Society. There had been many precedents in America and in Europe in previous years where so-called primitive peoples and their way of life were exhibited. The cultural chauvinism seems obvious now, but at the time there was a popular belief that tribal lifestyles were doomed by the social Darwinian triumph of western civilization. Well intentioned scientists and historians wanted to display and catalog the daily life of native peoples before they became extinct.

One of these preservationists was a young Cincinnati photographer named Enno Meyer. Meyer had photographed the Cree camp in 1895, and he spent many days capturing images of the Sicangu Sioux throughout the summer of 1896. Meyer's photographic record was donated to the Cincinnati Museum of Natural History in the 1980s, spurring a research project to identify the residents of the Rosebud Reservation that had made the trip to Ohio ninety years earlier.

Ironically, Meyer did not photograph the white cast members of the Cincinnati show. To date, no photograph of Maud Lee in Cincinnati—or anywhere else—has surfaced. In this case at least, the ephemeral popular culture represented by Maud has been lost and forgotten more quickly than the American Indian traditions that were feared to be in imminent danger of vanishing.

Maud was partnered in the show's Wild West skits with an eighteen-year-old from Illinois, Victor Anderson, who donned a buckskin coat to play the role of the frontier scout. Victor could shoot, but his main talent was knife throwing. Maud still had some lithograph posters left over from her days with S. F. Cody that read "Cody and Sister," so in order to

use up that supply, Victor Anderson was rechristened with a new professional name, Vic Cody.[11] Maud, who had acted like a big sister to Harry
Hill's son and like a mother to Vivian King, took Vic Cody under her
wing as her protégé.[12] Vic and Maud would reunite periodically between
1896 and 1902, but the relationship appears to have been more collegial
than romantic.

Vic Cody (Victor Anderson) and Lillian Cody (Maud Lee) were
the only billed white stars of the Cincinnati show, and they appeared
during its entire run from June 20 to September 6. The grounds where
they appeared included a stage area where other acts performed limited
engagements; those acts did not follow the Wild West theme. Each week,
there were two or three musical acts—singers or bands. The Cincinnati-
based John C. Weber Military Band appeared for several weeks, as did
Sie Hassan Ben Ali's Moorish acrobats. The Arabs, who were seasoned
vaudeville performers, were persuaded to mount horses and agreed to
stage a "Wild West meets Wild East" spectacle in which they and the
Sioux chased each other around the arena ring.[13]

Maud's health problems were apparently not an issue while she was
in Cincinnati. It is unlikely she could have performed the whole summer if she was using morphine, yet she completed the entire run of the
Cincinnati show and gained new friends among the Sicangu Sioux. An
October edition of the *New York Clipper* theatrical newspaper reported
that following the close of the Cincinnati run, Maud accepted an invita-
tion from the Sioux to visit the Rosebud Reservation. The *Clipper* also
mentioned that when the zoo show ended, she was presented with a Saint
Bernard puppy. This gift was a gesture of honor: nine years earlier, in
1887, at the conclusion of Buffalo Bill's successful engagement in London, Annie Oakley had been presented with a Saint Bernard puppy.[14]
Whoever gave the pet to Maud was bestowing upon her the same tribute
that had been awarded to Annie Oakley.

Maud's visit to the Rosebud Reservation is not documented, but there
is no reason to believe that the *Clipper*'s reporting was wrong. What Maud
would have seen at Rosebud would not have been anything like a large-
scale version of the Cincinnati encampment. The Rosebud Reservation was
among the most impoverished territories subject to federal Office of Indian

Affairs regulation. Sioux customs were repressed, particularly dance rites, religious celebrations, or anything related to warrior life. Children were taught in schools run by whites. The buffalo were gone, and food was scarce. The Indian Office doled out meager supplies of dried beef. If Maud had been in buoyant spirits on leaving Cincinnati, the dismal conditions in which she saw her new friends living must have saddened her.

Maud returned east in mid-October. The tent and arena show touring season was over, so she needed to book her shooting act into music halls. Being a low-salaried performer in small Mississippi Valley theaters presented logistical problems given the nature of her act, however. She could not afford the shipment costs to transport dozens or hundreds of delicate glass target balls from one town to the next. From her days with S. F. Cody, she knew the solution to this problem: she needed to make her own target balls at each stop.[15]

Homemade target balls were not made out of glass, but instead were molded from a heated liquid mixture of coal ash or pulverized fire clay, coal tar, and other ingredients.[16] It was possible to make them on an ordinary stove, although the stench was powerful. Once liquefied, the mixture could be poured into ball molds that folded together. After the outer shell cooled and hardened in the hollow mold, the remainder of the liquid inside the ball could be poured out and the mold opened. The result was called a "resin" or "composition" ball.

Initially, Maud did not know the technique for making her own target balls, but she was put in contact with someone who was willing to sell her the recipe. Maud later claimed that she had spent $100 to get the recipe—a small fortune in those days, and probably more than she had saved from her show in Cincinnati.[17] But she desperately needed those instructions in order to pursue her ambitions.

On October 24, 1896, Maud was arrested in Indianapolis, Indiana, for attempting to steal a $23 fur collar from the store of Boyd, Besten, and Langen. She was indicted by a county grand jury in mid-November.[18] It is not known what further legal actions resulted, but by early January, Maud's sharpshooting act was booked at the Casino Theater in Anderson, Indiana. Her first performance was on Monday evening, January 4, 1897. Before she was to take the stage again on Tuesday night, she toiled in an upstairs room at the theater, preparing her coal-tar mixture in a large pan on a stove.

Two other theater employees were in the room with her, a man and a woman. They started to engage in horseplay, chasing each other around the room. The woman, Ella Haverstick, was carrying a bucket of water. As she ran by Maud at the stove, she stumbled, and her water sloshed into Maud's pan of bubbling coal tar. There was an explosion as the two liquids reacted, and the hot tar burst out of the pan, spattering over Maud's arms and under her jaw. Maud screamed. People on the street nearby heard the explosion and thought it was a gun.

Someone called for a doctor to attend to Maud's burns; it was reported that she lost a great deal of flesh and suffered greatly throughout the night.[19] The attending physician, Dr. Willson, would have had only one drug in his kit that could have eased Maud's cries of pain from her burns: morphine.

During the summer of 1898, as S. F. Cody and his family were performing their Wild West show at Alexandra Palace, Auguste Gaudron and Harry Spencer were nearby testing what they called a "navigable balloon." Their airship, described at the time as porpoise-shaped, was sixty feet long and twenty-eight feet in diameter. At one end it was fitted with movable panels of sailcloth, which served as a rudder and a keel to stabilize and steer the craft. Slung underneath the balloon was a seven-foot-long wicker basket to carry the pilot. Attached to the basket was a $3^1/_3$-horsepower motor that turned a small aluminum propeller.

The craft's first flight took place on May 30, 1898. On hand as a guest witness was Captain Baden F. S. Baden-Powell of the Scots Guards. Baden-Powell, a noted expert on balloons, was secretary of the Aeronautical Society of Great Britain. The flight was less than a success. The motor and propeller proved too weak to drive the balloon against the wind, and the steering mechanism worked only during a brief calm. The balloon drifted eastward towards Epping Forest, unable to return to Alexandra Palace. It landed safely a few miles away.[20]

As the balloonists reevaluated their design that summer, they must have had serious discussions about the control and lift surfaces, or as they were then called, the "aeroplanes." Gaudron and Spencer were undoubt-edly familiar with the pioneering work on aeroplanes done by Lawrence Hargrave, an Australian engineer who, working alone, conducted experi-ments using tethered kite models. In the early 1890s, Hargrave had dis-covered that pairs of parallel aeroplanes, held apart by struts to form a box and having a bowed leading edge, provided great lift. His models came to be known as "Hargrave kites" or "scientific kites" and, more popularly, as "box kites."

Hargrave published his results in the mid-1890s, creating great inter-est in aeronautical societies around the world. The Hargrave design quickly proved itself to be superior to other kite or glider designs. In 1896 and 1897 he came to Britain to lecture on his experiments, which disseminated his findings to the ballooning community there. At about the same time, news reports in the popular press described a connected train of Hargrave kites being used to lift meteorological instruments to great heights by the Blue Hill Observatory at the U.S. Weather Bureau station near Boston.[21]

Alexandra Palace's resident balloonists must have discussed how Har-grave's aeroplane research might be applied to stabilize their airship. The palace grounds sat atop a high hill overlooking London, an ideal spot for kite flying. Hargrave's designs had been published in magazines, so it is possible that such kites might have been flown by hobbyists at Alexandra Palace in 1897–98. Whether or not S. F. Cody sat in on conversations with Gaudron and Spencer about Hargrave kites, or actually saw one being flown, he found himself intrigued. Moreover, he was already work-ing with lightweight stage scenery materials, and had his own tools and workspace. Cody also had an eager pair of assistants interested in kite flying: the boys Leon and Vivian.[22]

Earlier in 1898, during the months of February and March, Captain Baden-Powell gave a series of public lectures at the London Society for the Arts called "Kites, Their Theory and Practice."[23] Baden-Powell was interested in the military use of kites for observation, since balloons had proved problematic in windy conditions. Observation kites would need to lift heavy weights—cameras or even men—in order to be practical. In the mid-1890s, Baden-Powell had determined that only a train of multiple kites, rather than a single kite, would be able to provide enough lift to elevate the weights he had in mind.

Baden-Powell was not using Hargrave kites, however, and instead used a flat, somewhat hexagon-shaped design that he called the "Levitor." To add stability, he found it necessary to tether his kite train with two separate ground lines, spread out at wide angles from opposing sides of his kite. At points along these two ground lines, Baden-Powell hung secondary lines used to lift a payload basket. In 1895, his kite train rig raised a man in the basket several hundred feet into the air.

The idea of combining Hargrave-style box kite trains and Baden-Powell's goal of lifting heavy weights may have occurred to S. F. Cody very soon after his interest was piqued. His first concern was to improve the Hargrave kite's lateral stability. Whether he had help from Gaudron and Spencer is not known, but Cody's solution was to affix stabilizing panels jutting out at angles from the corners of the box struts. It seems to be more than a coincidence that Gaudron and Spencer were experimenting with aeroplane surfaces for their airship at the same time that Cody started to experiment with aeroplane surfaces in his kites.

Perhaps Cody initially intended simply to use his kites as another shooting target, just as he had employed model balloons in his arena show. Or, given his show business acumen, maybe he intended for the stable-flying kites to draw distant gawkers in to see his show—a trick that had long been used by circuses and fairs with promotional balloons. Whatever the reason, Cody's kite diversion came at a time when one would have thought he would be totally preoccupied in the production of his first, spectacular stage production.

By August 1898 Cody had nearly finished the scenery and stage properties he needed for his stage play, now titled *The Klondyke Nugget.* He tested the props on audiences from the stage at Alexandra Palace, including one scene in which a horse drops fifteen feet from a collapsing bridge. Cody must have remembered that the leaping horse stunt had been a crowd-pleaser back in his Forepaugh Circus days; and his fellow Forepaugh alumnus Doc Carver had likewise made jumping horses the centerpiece of some of his shows. Cody probably anticipated that some theatergoers would consider the scene to be mistreatment, and therefore he was prepared when representatives of the Society for the Prevention of Cruelty to Animals came to inspect the stunt. Cody was able to show them the deeply padded sub-platform onto which the horse fell, unharmed.[24]

Although it was a typical frontier melodrama in plot and character, *The Klondyke Nugget* was unique in its topical setting and some eye-popping stunts and stagecraft. The background for the story was the Klondike Gold Rush, which had begun in the summer of 1897. Gold deposits had been discovered in the Yukon Territory in 1896; when word reached the mining community and the outside world, a stampede of amateur miners descended on the Yukon via the Chilkoot Pass from Skagway, Alaska. They met immense hardships caused by the terrain and climate.

According to Cody's own version of his teen years, he had been a grubstake prospector in the Yukon in 1883–84, a time when fewer than 1,600 men were in Alaska or the Yukon Territory searching for a gold strike. No evidence has surfaced to back up this anecdote, and because Cody mythologized other events in his early years, there is no way to know whether he ever ranged that far north. At any rate, the claim of prospecting experience would have sounded authentic to the British theatergoing public, who already knew that Cody came from a colorful background. Cody also must have calculated that the Yukon setting, inasmuch as it was part of the Northwest Territories of the Dominion of Canada, would help convey the notion that there was still a "Wild West" within the British Empire.

The turgid plot of the five-act play involved a single father, Tom Lee, and his daughter, Rosie Lee. Rosie falls in love with a young prospector, Joe Smith, but another man, George Exelby, covets Rosie's attentions. Exelby has a criminal background, and proves himself a blackguard by plotting to jump a gold claim being worked by Joe, Rosie's brother Ted, and a friendly Indian named Waco. As the action plays out, the hero, Joe Smith, is framed for murder and sentenced to death. Just before he is executed, Waco, who had been thought dead, appears to refute the evidence against Joe, and Exelby is exposed as the villain who framed Joe. Exelby escapes to threaten Rosie and her father, but Waco and Joe arrive on the scene to kill Exelby and rescue the Lees.[25]

The audience likely cared little about the plot twists, simplistic characterizations, or stilted dialogue. *The Klondyke Nugget* offered inventive stage sets, abundant physical action, and auditory fireworks; few viewers went away disappointed. One scene took place in an underground mine shaft, with the stage designed to show actors both outside the mine opening and inside the shaft down below, as if the audience had a cross-sectional view through the earth. Another scene took place on a collapsing bridge over a mountain pass. Horses fell and crashed through windows, and there were hand-to-hand fist, knife, and gun fights. Cody cast himself as the villain Exelby, and Lela played Rosie, the leading lady. Her sons Edward, Leon, and Vivian portrayed the Indian Waco and Rosie's brothers.

Cody started advertising for theaters in October 1898. *The Klondyke Nugget* opened at St. George's Theatre in Walsall, near Birmingham, on December 5. Predictably, the first reviews of the play dismissed the acting and dialogue, but found it an enthralling experience all the same.[26] The Codys began a tour of provincial theaters that lasted until August 1899, when the play finally came to London. It was an immense success, and continued to tour Great Britain more or less continuously for the next four years. No one in the audience, and perhaps not even his costar and mate Lela herself, would have noticed that Cody chose to name the leading lady's character Rosie Lee, using the surname of his first wife, Maud Lee; or that he cast himself as the villain aiming to harm her.

As for Maud, a few years later she made a nonsensical claim that Buffalo Bill Cody had paid for her to have several stays in sanitariums throughout the 1890s, to help her recover from her afflictions. Yet William F. Cody had never met or employed Maud, and would have had little sympathy for someone who had appropriated his good name for professional purposes. So plainly Maud's claim is spurious.

Like many in the American theater world, Maud advertised a letter drop-box at the offices of the trade paper, the *New York Clipper*. S. F. Cody and performers in Britain used the *London Era* for the same purpose. Had they wanted to, S. F. Cody and Maud could have contacted each other through these letter drops. An American bit player in *The*

Klondyke Nugget once mentioned that the entire cast looked forward to receiving their copy of the *Clipper* from New York each week.

A question to ponder: if Maud did have multiple stays in sanitariums throughout the 1890s—an expense beyond the means of her parents—who sponsored that care for her, given that it was certainly not Buffalo Bill?[27]

Chapter 8

Kites and Buffalo

1899–1902

Samuel F. Cody's interest in the possibilities of Hargrave-style kites was piqued at Alexandra Palace in the months prior to the launch of his *Klondyke Nugget* tour in December 1898. What started as a diversion from the Codys' preparations for their first full-scale theatrical production soon became an obsession. S. F. Cody quickly went from single kites to multiple kite trains, and from smaller models to larger ones. By the time *The Klondyke Nugget* went on the road, he had developed a system to duplicate Baden-Powell's feat of lifting a man into the air in a basket. The Codys packed their kite equipment for transport along with their stage scenery.

As time and conditions allowed, Cody and his boys would launch their kites at the various theater towns they visited. For instance, in early April 1899, Cody sent aloft a kite train from the grounds of a wealthy landowner in Carlisle. A solid calendar of one-week engagements afforded Cody little time for further enhancements, however; that type of research required free time and access to a full workshop.

In October 1899, the Codys were playing an engagement at a theater in the town of St. Helens, near Liverpool—the same town in which S. F. Cody first appeared onstage upon his arrival in England in 1890. After the performance on the night of the thirteenth, when the theater was empty, a fire started that destroyed the entire structure. Cody lost all the scenery and props of his show . . . along with his kiting equipment. Incredibly, *The Klondyke Nugget* was able to reopen just two weeks later in Liverpool;[1] but Cody was unable to resume any kite testing until nearly a year later, in August 1900.[2]

While Cody was forced into kiting inactivity by the grind of his tour schedule and the loss of his equipment, Auguste Gaudron and Dr. Francis A. Barton were busy with their own aeronautical pursuits back at the Alexandra Palace. Barton was a medical doctor, specializing in obstetrics, but he had a keen interest in aviation and eventually was elected president of the Aeronautical Society of Great Britain. During the year 1900, Gaudron and Barton built several unmanned airship models, averaging sixteen feet in length, to test Barton's use of aeroplane panels for lift and vertical control.[3] Barton's goal was to build a series of airships with each successive model less dependent on gas bags for lift and more dependent on aeroplane frames.

Although the general public was skeptical and sometimes derisive of attempts to achieve controlled flight, the designers of these experimental craft knew that a breakthrough was only a matter of time. In America, the Wright Brothers were methodically testing aeroplane frame models and gliders; and the director of the Smithsonian Institution, Samuel P. Langley, was working on a man-carrying version of his successful unmanned aerodromes. In England, the Anglo-American inventor Hiram Maxim (the same man who had invented the machine gun mechanism used in Cody's Borchardt pistol) had tried to build a heavier-than-air craft in the early 1890s. He had abandoned that effort for lack of a powerful, lightweight engine, but still maintained an interest in aeronautical developments. Count Zeppelin constructed a new navigable balloon in which he conducted trials at Lake Constance on the Rhine in the summer of 1900. In France, Alberto Santos-Dumont was successfully flying dirigibles and was starting work on a heavier-than-air machine.

Most of these serious aeronautical engineers—with the exception of Santos-Dumont—preferred to keep details of their designs private. They were very much aware that their successes might one day translate into commercial production. Samuel F. Cody was similarly protective of the improvements he had made to the Hargrave box kite design. Cody applied for a patent on his kite design in November 1901, and at the same time contacted the British War Office to invite them to observe a demonstration. Cody was hoping that the War Office would be intrigued by the military advantages of his kites over the more unwieldy Baden-Powell design.

Although the War Office did send a few officers to observe Cody's ascensions, by the end of 1902, he had received only tepid interest from army commanders. Cody, who spent much of 1901 and 1902 still touring with *The Klondyke Nugget,* forged better relationships with weather researchers who were interested in taking atmospheric measurements by kite. In September 1902, a Cody kite reached an altitude of 14,000 feet, just short of the record set by the Blue Hill Observatory in the United States. In recognition of his efforts, Cody was made a Fellow of the Royal Meteorological Society in November 1902. He knew however that with additional money and better equipment, he could achieve even more spectacular results.

Maud Lee had a long road to recovery from the severe burns she received in Anderson, Indiana, in January 1897. If she had managed to stay free from morphine use before that incident, there is little doubt that the accident reintroduced her to the drug's comforts. Whether these events accelerated a preexisting mental illness cannot be stated with certainty, since Maud's whereabouts from 1897 to 1900 are unknown. The 1900 U.S. census, taken in early summer, found Maud living with her parents in Swedeland.[4] The fact that she was with her parents rather than touring may indicate that she was convalescing or mentally unfit to work. To the census taker she gave her occupation as "champion rifle shot." She was now twenty-eight years old.

In 1902 she once again found work in a Wild West show, but not as a featured star. In April 1902, Maud had made plans with another Wild West show rider named Lulu Mitchell to form a new stage sharpshooting act called "The Cody Sisters." Other than an announcement to that effect, however, there is no record that they ever appeared in any theater or show.[5] Instead, Maud found a job with a regional touring show headquartered in Paducah, Kentucky, called "Buckskin Bill's Wild West Company."

The Buckskin Bill show had originated two years earlier, in 1900. During its first two years it was owned and managed by an experienced circus man, Fletcher Terrell.[6] Terrell and his three brothers had operated the Terrell Brothers Circus earlier in the 1890s, confining their tours to railway towns in the Ohio Valley. The same area became the main route of the Buckskin Bill tour.

Fletcher Terrell avoided the danger of employing a high-priced, prima donna frontiersman to front his show by instead creating the fictional character of "Buckskin Bill." This character was modeled on a younger version of Buffalo Bill Cody, complete with buckskin coat and leggings, long hair, and mustache. Terrell's shrewd ploy was to create an instantly recognizable character into which he could insert malleable, little-known cowboy performers. The fact that several real-life western characters, as well as dime-novel fiction and minor vaudeville acts, had used the name Buckskin Bill helped to confuse the public. W. F. "Harry" Brandon played "Buckskin Bill" in 1900, but in 1901 the honor was bestowed—at a reasonable salary—on Maud's former protégé, Victor F. Cody.[7]

Before the show's 1902 season, the Terrells sold their share of the outfit to W. E. Allott and J. C. O'Brien.[8] The new investors expanded the show to a troupe of 450, making it the second largest Wild West show in America in 1902. (The largest, Buffalo Bill's Wild West, toured the West Coast that year before leaving for Europe.) Among the troupe were Sie Hassan Ben Ali's Moorish acrobats and a team of Cossack riders led by Alexis Georgian. Vic Cody once again started the 1902 season in the lead role of Buckskin Bill, but later in the year he was replaced by another little-known rifleman, "Cherokee Bill" Cahoon.[9] It may be that Vic Cody helped Maud to join the troupe when it first formed in April. The show started its season in Paducah on May 3, 1902, then headed northeast.

On May 15, less than two weeks into the tour, the Buckskin Bill show played a date in Russellville, Kentucky. Maud was performing as a "rough rider" under her stage name Lillian Cody. During her segment, she was thrown from her horse and fell violently to the ground. She screamed in pain as she lay in the dirt and when she was carried out of the arena. Doctors were summoned and later determined that she had a dislocated hip in addition to several bruises. The pain associated with such a hip injury is enormous; very likely the immediate treatment Maud received was sedation with an opiate, such as morphine. Once she was sedated, doctors then would have had to manipulate her femur back into her pelvis.[10]

The recommended recovery period for such an injury is two or three months, but Maud could not afford to be inactive. The Buckskin Bill show had long since moved on from Russellville, and she did not rejoin it. Instead, in late June, after barely a month of recuperation, Maud responded to an ad in a theatrical newspaper calling for cast members for a new Wild West show forming in Missoula, Montana, called "The Great Buffalo and Wild West Shows United."

The new show was the ambition of two Missoula brothers, William A. and Frank E. Simons. They asked an experienced show promoter, George L. Hutchin, to serve as a third director. The focus of the show was built around thirty American bison taken from the last remaining wild herd in the United States, owned by Charles Allard, Jr., and Michel Pablo and contained on their ranch at Wild Horse Island on the Flathead Indian Reservation. The population of American bison had reached a nadir of about six hundred in the mid-1880s.[11] By 1902, careful protection (through both private and public efforts) had increased the buffalo population by several hundred, but the popular conception was that the bison was still on the brink of extinction.

The Simons hoped that people would pay to see live examples of this remnant of the American frontier before it disappeared. As if to emphasize that the bison were the true stars of the show, the human performers received only minimal billing, and none was given the salary that a headline star usually collected. Hutchin was quoted as saying, "This is not a ONE MAN SHOW. Every actor and performer is a star."[12]

This philosophy apparently did not sit well with Victor Cody, who, like Maud, had left the Buckskin Bill show and answered the initial casting call from Montana. Although he arrived for the assembly of the troupe in mid-June, by the time the show opened in the first week of July 1902, Vic Cody was nowhere to be seen.[13] That left the honor of leading the traditional opening parade through each town on the route to Lillian Cody, that is, Maud Lee.[14] She received top billing over even the show's master of ceremonies, veteran Al G. Barnes.

The show was previewed to a Missoula audience on July 2. Though the day was rainy and cold, more than three thousand people turned out to cheer the hometown production. The Missoula newspapers gave the show glowing reviews, but also noted that there were long delays between the acts. The *Daily Missoulian* hailed Lillian Cody as "an exceedingly clever shot and rider." Of the show itself, the reporter said, "The many thrilling features of the show brought forth the delight of the audience—stage coach chase, horse thief capture and lynching, pony express and broncho busting—being as perfect as could be conceived." Unfortunately, this first report, undoubtedly written as a favor to the Simons brothers by the hometown paper, was one of the few positive reviews the show would garner.[15]

The production was due to play in Butte, Montana, the next day, but wet canvas and uncooperative buffalo made loading the show train difficult. The late departure mattered little, because five inches of snow (in July!) cancelled the Butte engagement. The train continued on to Helena, but so did the rain and snow, limiting the stay there to just two well-attended shows over three days. The next stop was Billings, and the reception there was chilly in more ways than one.

At Billings, the reviews were scathing: "The Buffalo and Wild West aggregation which exhibited in the city yesterday was about as complete a failure as one could imagine from the viewpoint of a show. . . . There is not a meritorious act in the entire programme. . . . The company itself is sufficient to condemn it in the eyes of the public when it appears in a street parade."[16]

From Montana, the show traveled on to Minnesota, Wisconsin, and Illinois, hitting rain on forty-one of forty-five days. Although the tent kept the ring area and folding chairs dry, the weather discouraged large audiences from attending. The rain made the process of packing and unpacking the canvas, animals, and equipment an arduous task. The prolonged grind of travel in bad weather started to take a toll on the livestock. By the end of August, the show stopped dead in Galena, Illinois, because Charles Allard, Jr., the rancher who had supplied the buffalo, insisted that the buffalo and horses needed a rest on risk of death. Meanwhile, the Simons brothers and George Hutchin bickered over their mounting financial losses.

Hutchin urged the show back on tour and steered it toward Iowa. While on the way to Waterloo, Iowa, another of the bison expired, bringing casualties in the herd up to six—and those six had been the largest, most impressive specimens in the show.[17] The cast of the show must have observed the misery of the bison, and the effect that rail travel had on them, and wondered if they were hastening the extinction of the species. As each animal died, so did a little piece of the real Wild West.

While stopped in Waterloo on September 4, Hutchin was forced by debt to sell four more buffalo from their dwindling corral. Yet even that measure fell short; a group of the show's investors went to the local justice of the peace and persuaded him to issue an order to sell the gray horse team used to pull the wagon of the show's brass band. At Waterloo, one of the buffalo escaped and could not be found before the train had to leave.

From Waterloo, the show was able to fulfill a performance date in Cedar Falls, Iowa. It then moved to Marshalltown, Iowa, and floundered there for good. Many of the cast members had already deserted in the week leading up to the show's arrival in Marshalltown.[18] George L. Hutchin tried to resolve the problem that the stranded outfit presented to the good citizens of Marshalltown by promising to make the town the new winter headquarters of the show. Few believed him after they saw its salvageable remnants being loaded onto a train back to Montana. Charles Allard, Jr.—accompanied by his remaining bison and by the Bitterroot Salish, Pend d'Oreille, and Kootenai Indians that had traveled with the show—returned to their homes on the Flathead Reservation.

As they headed home, the Montana-born roustabouts swore to themselves they would never get involved with show business again.[19] Maud Lee, however, as a professional performer, faced the task of finding her next job—and between Marshalltown and Missoula there were few prospects. She needed to get back to a big city, where she could touch base with the informal network of theatrical contacts. She must have been exhausted and depressed from this recent tour, and very likely was still taking morphine to ease her hip injury.

CHAPTER 9

CAST TO THE WIND

1903

As S. F. Cody entered the year 1903, he became convinced that scientific kites offered many applications beyond military observation. One application had already been tested: in December 1901 the Marconi Company successfully used Baden-Powell Levitor-style kites as radio transmitters to receive the first transatlantic radio signal sent from England to Newfoundland. Guglielmo Marconi also believed that kite-antennas were ideal for ship-to-shore and ship-to-ship radio signaling.

Although Cody's overtures to the War Office had been met with lukewarm interest, in February 1903 he decided to approach the British Admiralty directly to promote naval kite applications. In March, the admiralty sent observers to see Cody fly kites of different sizes at Woolwich; one of these tests was done with his large man-lifting system. These demonstrations proved interesting enough for the admiralty to request additional trials at sea. At Portsmouth, Cody showed navy officials how the kites could be used for radio transmissions and observation. Aboard the HMS *Sea Horse,* Cody himself ascended in the basket-chair from the deck of the ship.

Despite Cody's impressive demonstrations, when he asked for contracts to supply kites to the military and to be retained as a consultant, he was flatly turned down by the Office of the Director General of Ordnance. Cody had been prepared to give up performing to devote himself to aeronautics full time. Instead, when *The Klondyke Nugget* returned from its spring tour, Cody reestablished his headquarters at Alexandra Palace and gave shows there throughout the summer. He now used the spacious Old Banqueting Hall of the palace as his kite-storage facility, which made for an anachronistic contrast between the faux medieval banners on the walls and the cubist geometric shapes of Cody's devices dangling from the ceiling. It must have been small consolation to Cody that the admiralty, perhaps in defiance of the Ordnance Office, ordered four kite systems from him.

Britain's small but active aeronautical community was quick to recognize the value of Cody's continued work on kites. On June 24, 1903, he was made a member of the Aeronautical Society of Great Britain. By 1903, Cody had met and likely conversed with many of its members: Francis A. Barton, B. F. S. Baden-Powell, Patrick Alexander, Thomas W. K. Clarke, Hiram Maxim, and others. Many of these men had been in personal contact with the premier aviation figures of the era: Octave Chanute, Otto Lilienthal (before his death in 1896), Samuel P. Langley, Ferdinand von Zeppelin, and even the unheralded Wright Brothers of Ohio. Cody's acceptance into the Aeronautical Society placed him firmly within a group that was keenly aware of aviation developments happening around the world.[1]

Cody's personal agenda to gain support for his kiting efforts now was matched by a new determination by the Aeronautical Society to sponsor public events. The day after Cody was accepted into the society, it promoted a high-altitude kite-flying competition at Findon, near Worthing. Cody was one of the four competitors.[2] Cody also seemed to have had a hand in organizing an aeronautical exhibition a few months later, in September 1903. The displays were housed in his personal "hanger," the Old Banqueting Hall at Alexandra Palace.

Though competitions and exhibitions raised public interest, in Cody's mind some type of dramatic action was needed to reengage the interest of the British military establishment. One idea he suggested in an article in *Pearson's Magazine* (published in both Great Britain and America) was to use kite power to pull sleds across the ice on polar expeditions.[3] In fact, one British newspaper interview suggested that Cody considered undertaking an arctic expedition himself.[4] He was intrepid enough to have followed through on such an adventure, but lacked the wherewithal even to get started.

Another sensational endeavor came to Cody's mind, however. It was a stunt that never ceased to capture the imagination of the British public and garner headlines around the world: crossing the English Channel. Cody had only to look at recent publicity given to balloonists who accomplished that feat: Percy Spencer (elder brother of Harry Spencer, Gaudron's partner) had crossed the English Channel in a balloon in 1898. A generation earlier, in 1875, American Paul Boyton had won enormous acclaim for making the crossing in a buoyant rubber suit. His accomplishment was bettered a few months later by Matthew Webb, the first person to swim unaided across the channel.

The most recent channel-crossing adventure Cody could look to was the balloon flight by his friend Auguste Gaudron and Dr. Francis Barton in August 1902. Gaudron and Barton barely made land at an isolated beach ten miles from Calais, which triggered sensational headlines, such as "Balloon in Channel Gale—Stranded on Quicksands."[5]

Cody proposed a stunt that combined air and sea navigation—he announced a plan to sail across the English Channel in a kite-powered boat. He chose for his vessel a lightweight folding twelve-foot Berthon boat.[6] The Berthon had proven itself in the Boer War as a valuable troop craft for fording rivers, and was also touted as a storage-saving life raft for large passenger liners. On October 9, 1903, Cody tested the motive power of the kite-boat at Dover. The next day he made an attempt at the channel crossing, but after three hours the trip was aborted after slack winds forced the kite's tether line into the water.

He had to wait four weeks before his schedule and weather conditions allowed another attempt. On November 5, he found the winds were blowing favorably from Calais to Dover, so the boat and kites were loaded onto a mail packet to France. He began his trip in the evening, just before 8 P.M.. Cody made good time for the first few miles, but halfway through the crossing the wind fell and he had to haul the kite in. He drifted with the tide until the wind picked up, then relaunched the kite. As he neared Dover, he had to retrieve and relaunch the kite a second time, but finally landed after an eleven-hour journey.[7]

Cody's channel crossing was lauded with fanfare and global press coverage. He capitalized on the publicity with several public lectures describing the journey. Undoubtedly, his primary intent was to pique the interest of the stodgy British military establishment; however, it so happened that Cody had written a new play *Viva* that debuted a month after his voyage. The acclaim Cody had received was wasted on *Viva,* which failed to recapture the magic appeal of *The Klondyke Nugget.*

At any rate, between his channel-crossing venture and his contacts in the aeronautical community, Cody was emboldened to approach a different arm of the military hierarchy: the Balloon Factory unit of the Royal Engineers. At the close of 1903, he wrote to the commander at Aldershot, inviting him to view a demonstration of the capabilities of his kites. Ultimately, it was through the Balloon Factory that Cody gained the foothold he needed to further his aviation ambitions.

John A. Forepaugh, a nephew of circus owner Adam Forepaugh, had married a young opera singer from California named Luella Jordan. John was active in Philadelphia politics and owned a casino and theater in that city. After John died in 1895, Luella remarried a man named George Fish, one of John Forepaugh's pallbearers. In 1897, Luella and George authored a stage adaptation of Robert Louis Stevenson's *Strange Case of Dr. Jekyll and Mr. Hyde,* a harrowing account of a personality disorder caused by addictive drugs. After staging this production in the theater she had inherited from John, Luella and George sold the theater and decided to compete against other members of the Forepaugh family by going into the tent-show business.

The Luella Forepaugh-Fish Wild West Show opened its first season at its headquarters city of St. Louis, Missouri, on April 17, 1903. The show featured male sharpshooter "Cherokee Bill" Cahoon, who had the previous year starred as Buckskin Bill in the Buckskin Bill Wild West Show. The female shooter-equestrienne was "Lone Star May" Mackey. Also in the troupe were Lakota Sioux from the Pine Ridge Reservation, a circus menagerie, ex-troopers from the U.S. Cavalry, Arab acrobats, and Alexis Georgian's crew of fourteen Russian Cossack horsemen.[8]

The show had a miserable month of May, missing more than a third of its performance dates due to rain as it traveled through Missouri, Kansas, Nebraska, and Iowa. As it limped through Minnesota and Wisconsin in June, the show's debts mounted and it deferred payment of the performers' salaries. To their disgust, Luella and George discovered that their own employees had been skimming the admittance receipts. They decided to sell the show after it reached Janesville, Wisconsin, at the end of July.

The show was bought on August 1 by F. J. Walker, the head of the lithography company that provided the show with its posters. Walker retained John Barton as the manager, and Barton implored the performers who had been stranded to accept new contracts. Many of them had already departed, leaving the show with many holes in its cast. For instance, it needed a new lady sharpshooter. The head of the Russian Cossacks, Alexis Georgian, was recruited to go south from Janesville to Chicago, in order to find an "Annie Oakley."

In early August 1903, Maud Lee was mired in poverty and, most likely, morphine addiction. She was living in Chicago, in a three-story brick tenement house in the roughest area of the city, a neighborhood known as the Armory. The large building of apartments where she resided was known as the center of Chicago's Assyrian colony, but she roomed in an apartment belonging to an older African American man named Charles Curtis. The nature of the relationship between Charles Curtis and Maud Lee is not known, but Maud later did mention that Curtis treated her with kindness when she needed help.

By her own account, Maud signed a contract to appear with the Luella Forepaugh-Fish Wild West Show and allowed her kit of guns, tackle, and target-ball molds to be sent up to Wisconsin. Apparently, she was the "Annie Oakley" that Alexis Georgian had found on his mission to Chicago (the two knew each other from the 1902 Buckskin Bill company). It is unlikely that Maud was given any advance money. As it turned out, Maud herself was never able to join the show. This was not a great loss of opportunity for her, because the revived Forepaugh-Fish Wild West Show managed to perform just a few more dates before closing down for good on August 12.

On the morning of Saturday, August 8, the kindness of Charles Curtis turned to outrage when he discovered his fine pair of trousers was missing. He concluded that Maud had stolen them to sell for drug money, and went to find a police officer. He signed a complaint against Maud that was received by Officer John J. Mulcahy. When Mulcahy later came to arrest Maud, Curtis had second thoughts about pressing his complaint, but Maud was intoxicated with drink or drugs and unruly, so Mulcahy arrested her on charges of disorderly conduct. She was brought to the Harrison Street Police Court two blocks away from the tenement, where she was placed in a cell to await an appearance before a judge on the following Monday morning.

Maud was kept in a downstairs women's cell supervised by police matrons. While Maud spent Saturday deprived of her intoxicants, Officer Mulcahy told others of the things he had heard while arresting Maud. Mulcahy said that Maud had claimed to be Annie Oakley.[9] However, when Maud was brought before the Police Court she gave her name as Lillian Cody. Whether Maud actually told Mulcahy that she was Annie Oakley, or whether Mulcahy misinterpreted what he heard during the fracas of her arrest, is a crucial question. Curtis or Maud could have declared that she had been a champion sharpshooter in Cody's Wild West show—or *like* Annie Oakley (or *an* Annie Oakley) rather than Annie Oakley herself; but that nuance might have escaped Mulcahy.[10]

About 11:00 P.M. Saturday evening, the evening shift of police matrons came on duty, including a matron named Anna Murphy. Anna was on duty at about midnight, when three "foreign" gentlemen, at least one of whom was English and all of whom were in town for a newspaper convention, came to the Police Court in search of a promising story. They had heard that Annie Oakley was in the lockup, and they obtained permission to visit her cell. For nearly five hours they listened to Maud tell her life story and peppered her with questions. During two of those hours, between 2:00 and 4:00 A.M., Anna Murphy listened to those conversations.

The length of time the three foreign newspapermen spent with Maud indicates they found her story puzzling and compelling, but there is no record that any of them ever published a story about Maud Lee or Annie Oakley. Only Anna Murphy's account of those conversations survives, and based on Murphy's testimony, Maud told a story that mixed delusion and false claims with revealing truths, related in a halting style that made it difficult for ordinary listeners to discern fact from fiction.

Boiled down, Anna Murphy's version of what Maud said is as follows: Maud had been married to Sam Cody, who was Buffalo Bill's son. Maud had been a sharpshooter in Buffalo Bill Cody's show before another female shooter took over her role as "Annie Oakley." She had performed in England and in Europe, and had done so when King Edward was in the audience. She had a baby boy named Vivian, who was being cared for in the Dakotas by her mother-in-law, the wife of Buffalo Bill Cody. She had an argument with the woman who took over the role of "Annie Oakley," and the other woman told her it was Maud's own fault for losing the role because of her drug habit. The other "Annie Oakley" woman told Maud she was divorcing her own husband in order to marry Sam Cody, and Maud relented because she thought this other woman was better for Sam Cody than she was.

Maud said that Sam Cody was now dead. She admitted to having been addicted to morphine in the past, but said that Bill Cody had sent her to different institutions and sanatoriums to be cured of the habit. She said that during one period when she was not using morphine, she had been with a Wild West show that advertised her as "Annie Oakley," and that Buffalo Bill's business agents had threatened to have her arrested if she used that name.

Anna Murphy described Maud as twenty-nine or thirty years old, with dark hair and dark eyes. Her clothes were unclean and her physical appearance was pale and dissipated, but she had cleaned her face and combed her hair. Murphy also said that Maud appeared to be sober by the time the former came on duty, more than fourteen hours after Maud had first been brought in. She had no visible skin sores, but did wrap a cloth around her neck under her jaw. Murphy also said that she thought Maud looked like someone dissipated more by drink than by drugs.

Maud told Murphy that she thought she would not be prosecuted, because Charles Curtis was a kind man and had given her shelter when she was destitute. Maud said that she would return the missing trousers that had made Curtis upset. Maud was apparently unaware that she would not be facing a charge of stealing the trousers—she would be charged for being intoxicated and causing a disturbance.

Charles Curtis did come to visit Maud on Sunday morning, and brought her coffee and a sandwich. Before noon on Sunday, four or five more reporters who had heard that "Annie Oakley" was in jail came by her cell to talk. On Monday morning at about 9:30, Maud was brought upstairs to the court of Police Magistrate John R. Caverly by Officer Mulcahy, the man who had arrested her. Maud's name was announced in the court records as Lillian Cody. Mulcahy stepped forward before Caverly and said, "This woman claims to be Annie Oakley, the famous woman rifle shot. She is full of disease and morphine and has been living with a colored man, Charles Curtis, on Sherman Street, I think. She is accused by him of stealing a pair of trousers."

Caverly discounted Mulcahy's statement of the charge, since Curtis was not present to press his complaint. Caverly did, however, take notice of the recorded disorderly conduct charge and of Maud's condition. He noted that she was poorly clad, dirty, and disheveled. She hung her head down, but he saw no sores on her face. Maud did not speak a word while in court.

Caverly believed that her appearance was consistent with the aftermath of intoxication that led to the disorderly conduct charge. He fined her $25, which he was probably well aware she would be unable to pay. He judged that a period of confinement in a house of corrections, which would be enforced in lieu of the fine payment, would prompt Maud to change her bad habits and improve her physical condition.

Maud was sent back downstairs to the holding cell to await a bus that would take her to Chicago's Bridewell prison. While she was waiting, several more reporters asked for permission to speak to her. Matron Murphy led them down to her cell. Among them was a reporter for William Randolph Hearst's *Chicago American* named George W. Pratt. Pratt thought he had an advantage in determining whether or not Maud was the real Annie Oakley—he had seen Buffalo Bill's Wild West when it was at the Chicago Columbian Exposition in 1893 not just once, but seven times.

Over the next hour before the bus arrived, Pratt and the others heard Maud tell her jumbled life story again. To make their noon deadlines, the reporters had little time to file their stories or call in their notes by telephone. Since they all heard the same details, none of the men believed that he had scooped the story. Pratt called in his notes in time to meet the deadline, and the *American* editors wrote up the story and published it in the 5:00 P.M. edition. The *American*'s story was then distributed through national news services, and by Tuesday had appeared in dozens of papers. The headline varied in each city, but the datelined story remained the same:

Annie Oakley, daughter-in-law of "Buffalo Bill" and the most famous woman rifle shot in the world, lies today in a cell in the Harrison Street station under a Bridewell sentence for stealing the trousers of a negro in order to get money with which to buy cocaine.

[This was] the woman for whose spectacular marksmanship King Edward himself once led the applause in the courtyard of Buckingham Palace.

When arrested Saturday on the complaint of Charles Curtis, a negro, she was living at 140 Sherman Street. She gave the name of Elizabeth Cody, but it occurred to no one to connect her with Colonel Cody's famous daughter-in-law. Yesterday, however, when brought before Judge Caverly she admitted her guilt.

"I plead guilty your honor, but I hope you will have pity on me," she begged. "An uncontrollable appetite for drugs has brought me here. I began the use of it years ago to steady me under the strain of the life I was living, and now it has lost me everything. Please give me a chance to pull myself together."

The striking beauty of the woman whom the crowds at the World's Fair admired is now entirely gone. Although she is but twenty-eight years old, she looks almost forty. Hers in fact is one of the extreme cases which have come up in the Harrison Street police court.

Today she will be taken to the Bridewell to serve out a sentence of $45 [sic] and costs. "A good long stay in the Bridewell will do you good," said the court.

The prisoner's husband, Sam Cody, died in England. Their son, Vivian, is now with Colonel Cody at the latter's ranch on the North Platte. The mother left "Buffalo Bill" two years ago and since has been drifting about the country with stray shows.[11]

Pratt had done nothing to verify Maud's story, and many of his details were wrong. In addition, either Pratt or his editors decided to embellish some of the facts. No one checked to see whether William F. Cody had a son named Sam or a daughter-in-law; he had neither. Anyone in show business could have informed them that Annie Oakley was married to a man named Frank Butler.

Among the details that were incorrect Maud was sentenced for disorderly conduct, not stealing; she had admitted to former morphine use, not current cocaine use. She never admitted that she was currently taking drugs or that she had stolen the trousers for drug money—though that may have been true. She never begged nor pleaded before Justice Caverly—she was silent the whole time she was being sentenced. She was not twenty-eight years old, but thirty-one. Twenty-eight would have been impossibly young for the real Annie Oakley, who was forty-three. Finally, Maud was fined $25, not $45.

The truth did not matter. Pratt and his editors had a juicy story about the fall of a public figure, a scandal highlighted by a suggestion of racial miscegenation and the terrors of drug addiction. It was a story custom-made for the lax yellow journalism ethical standards of the newspaper industry at the time, a style promoted by the *American*'s owner, William Randolph Hearst.

When the story was disseminated through the news services, some papers decided not to print it, realizing it was suspect. Those that did print it composed their own headlines. The story appeared under such banners as

"Woman Dead Shot Is Fined; Annie Oakley, Who Says King
Edward Applauded Her Marksmanship, Admits a Theft"
"Annie Oakley's Disgrace: Stole Trousers of Negro to Get
Cocaine"
"Annie Oakley's Fall: Once Famous Woman Now in Chicago Jail,
a Cocaine Victim"
"Annie Oakley Is a Wreck"
"Annie Oakley in Prison Cell"
"Annie Oakley's Plight: Famous Performer Arrested in Chicago
for Robbery"
"Dope Caused Her Downfall; Annie Oakley Is Now in Jail"[12]

Maud Lee, having been transported to Chicago's dismal Bridewell Prison on Monday afternoon, was oblivious to the fact that her plight was being reported in newspapers across the country, and that it was being done so under the name of Annie Oakley. Almost immediately, however, the news reports reached the Atlantic Highlands, New Jersey, home of Frank and Annie Butler—the real Annie Oakley. The Butlers' first response upon hearing one of these reports was to demand a retraction; but as the magnitude of the broadcast fabrication was revealed, they decided to respond with litigation. In an unprecedented legal action, the expert rifle shots initiated libel suits against more than fifty newspapers across the United States.

Opening of the roller-skating rink at the Olympia National Agricultural
Hall, London, April 1890. S. F. Cody and Maud Lee appeared at the rink in
December 1890, in a burlesque of Buffalo Bill's Wild West. (*Pall Mall Budget,*
April 24, 1890)

The Arab riders of the 1890 French Exhibition "Wild East" program, in which S. F. Cody and Maud Lee performed as a Chasseur d'Afrique officer and a *cantinière*. The evocative scenery was painted on canvas backdrops at the Earls Court outdoor arena. (Engraving from the *Penny Illustrated,* May 31, 1890, p. 10)

John Robinson Whitley, the producer of the four Earls Court national exhibitions: American (1887, the exhibition that introduced Buffalo Bill to England), Italian (1888), French (1890), and German (1891). (From *John R. Whitley: A Sketch of His Life and Work*)

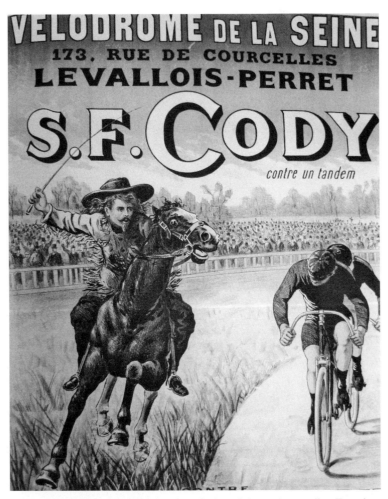

A poster advertising the match between Cody and the tandem cyclists Fournier and Gaby. Although Cody's horse-versus-bicycle races were initially popular, they soon drew derision and blacklist threats from organized cycling unions fearful of a gambling scandal. (Author's collection)

Annie Oakley was presented with a Saint Bernard pup, Sir Ralph, in England in 1887. The same gift was made to Maud Lee in 1896 after the Cincinnati Zoo exhibition closed. Although the matching gifts may be coincidence, the common denominator was Major John M. Burke, associate of Buffalo Bill and advisor to the Cincinnati exhibition. (Photo P.69.1190, Original Buffalo Bill Museum Collection, Buffalo Bill Historical Center, Cody, Wyoming)

S. F. Cody, a portrait likely taken during the *Klondyke Nugget* years,
1898–1904. Due to his melodramatic scenes in the arena and onstage, Cody
developed a talent for striking theatrical poses. (Photo LC-DIG-ggbain-03572,
George Grantham Bain Collection, Prints and Photographs Division, Library of
Congress)

A poster advertising Cody's play, *The Klondyke Nugget*. Cody worked on the scenery in a workshop at Alexandra Palace. (Photo by Sara Boesser; image property of Alaska State Museum)

Women riders of Pawnee Bill's Historic Wild West show. Although Maud Lee is not known to have toured with this troupe, the trick riding depicted in this poster is representative of the stunts she would have performed. Insurance companies determined that arena show riding was one of the riskiest professions. (Photo LC-USZ62–21486, Prints and Photographs Division, Library of Congress)

The Great **Buffalo Wild** West Shows **United**

WATERLOO } **Sept. 3**
WEDNESDAY }

Greater and Grander Than Ever Before

Exhibiting the only herd of Wild Buffalo on earth
Noted Chiefs, Warriors, Bucks, Squaws and
Papooses. Congress of Rough Riders of the world.

See the Cow Boys and Cow Girls. the Gauchos of
South America. the Bedouins, Cossacks and Vaqueros
of Old Mexico. The famous and historical Deadwood Stage Coach attacked and repulsed.

Fancy Rifle Shooting by the Crack Shots of the
world. Champion Ropers and Lariat Jugglers.

Big As It Is Good

1000 Men, Horses and Trained Animals used in
this production of Border Life.

A vivid reflex of Savagery opposing the climaxing
of civilization in the far west.

Two Exhibitions Daily

RAIN OR SHINE, 2 and 8 P. M.

A Grand Calvacade headed by Lillian Cody and
Prof. Wheeler's Famous Concert Cowboy Band
will parade. The pageant is a brilliant spectacle of
barbaric and savage splendor.

EXCURSIONS on all RAILROADS

The Great Buffalo and Wild West Shows United comes to Waterloo, Iowa . . .
and is stranded there. The near extinction of the American bison was seemingly
hastened by livestock losses during this tour. Note that Lillian Cody (Maud
Lee) led the parade. (*Iowa State Reporter,* August 29, 1902)

A. E. GAUDRON

Aeronaut & Balloon Manufacturer

Gold Medal, Aero Club Exhibition, 1906

Balloons of every description

In Silk, Cambric, Goldbeater Skin, Gutta Percha, etc.

Constructor and Aeronaut of the Barton
Airship Balloon

CONSTRUCTION OF AEROPLANES TO ORDER

Every assistance rendered to Experimenters in " Heavier
than Air " Inventions.

Light Motors supplied.   Large Works and Grounds
for Experiments at disposal.

BALLOON ASCENTS ARRANGED AT SHORT NOTICE

in Town or Country.   Seats may be booked in advance.

Special Ascents for Scientific Purposes, Cross
Channel Trips, etc.

Model Balloons built for Inventors.   Experiments
conducted.

HYDROGEN GAS APPARATUS

Advertising Balloons made to any Shape

Terms and Particulars—

A. E. GAUDRON Alexandra Palace
LONDON, N.

Telegrams—''GAUDRON, LONDON.''

A 1907 advertisement for balloonist Auguste Gaudron. Note that his services include construction of "aeroplanes," by which he likely meant control surfaces, not airplanes. Also note his model balloons, which may have been the means by which Cody and Gaudron were introduced. (From the *Pocket Book of Aeronautics,* 1907).

133

Cody standing astride his King Kite, the largest box kite in the world. The kite was about thirty feet long, and contained 675 square feet of canvas. Photo taken in 1901 or 1902. (Samuel F. Cody Collection, Drachen Foundation)

The British Army airship *Nulli Secundus,* built and piloted with the assistance of S. F. Cody. Note the aeroplane control surfaces and rudder. (Courtesy of Jean Roberts)

Cody, wearing yellow oilskins, poses in the Berthon boat he used to cross the English Channel. The photo was taken in the hall at the Alexandra Palace, November 1903. Note the board to the right with pictures of Cody's kiting projects. (Samuel F. Cody Collection, Drachen Foundation)

Cody and members of the Royal Engineers view the occupant of a wicker basket underneath a large Extended Wing Carrier War Kite, about forty feet off the ground. Cody usually used trains of kites for man-lifting. Taken at Farnborough, circa 1906. (Samuel F. Cody Collection, Drachen Foundation)

Cody standing next to his hastily assembled plane, which won the 1912 Military Aeroplane Trials at Lark Hill, Wiltshire. (Courtesy of Jean Roberts)

A postcard of S. F. Cody's funeral procession. He was buried with military honors. Maud Lee's parents made an immediate claim on his estate. (Courtesy of Jean Roberts)

The Women's Convalescent Building, Norristown State Hospital. Maud Lee spent the last forty-one years of her life confined to the grounds of this asylum. (Reprinted from the 1906 *Annual Report of the Norristown State Hospital*)

CHAPTER 10

HEIGHTS AND DEPTHS

1904–1908

On December 17, 1903, the Wright Flyer, a man-carrying powered craft with wings inspired by Hargrave kites, moved along a sixty-foot track on the sands at Kitty Hawk, North Carolina, and lifted into the air. It rose ten feet off the ground and traversed 120 feet along a straight path, with a few abrupt drifts up and down, before dipping to the earth. The Wright brothers, Wilbur and Orville, made three further flights that day before a strong wind damaged their machine. These first flights have since earned the status of a pivotal moment in aviation history. In contemporary aeronautical circles around the world, however, the brief news item reporting their success was assessed as just another step forward to a goal not yet fully achieved: prolonged and controlled powered flight.

From the perspective of these aeronautical researchers, the Wrights' achievement was an expected progression based on earlier developments, such as Otto Lilienthal's glider flights of 1891–96; Octave Chanute's glider tests using aeroplane wings; and the powered, unmanned aeroplane craft flights of Hiram Maxim, Clement Ader, and Samuel P. Langley. Because there were few witnesses to the Kitty Hawk flights, it was not apparent from the first accounts whether the Wright Flyer had been assisted into the air or whether the pilot had any control over the craft. The Wrights themselves discouraged publicity because they wanted to preserve the privacy they felt they needed to make proprietary, patentable refinements to their design.

In England, those in the know realized that the Wrights were in the vanguard of aeronautical research. During 1904 the Wrights made only nonpublicized flights with a new flyer at Dayton, Ohio. However, the usually secretive brothers accepted visits from two English members of the Aeronautical Society of Great Britain. The Wrights, frustrated by lack of support from the American government, were interested in seeing if the British would fund their efforts. One of the visitors was the personable aviation enthusiast Patrick Alexander, who by that time was a friend to Samuel F. Cody. The other was Colonel John Edward Capper, superintendent of the British Army Balloon Factory at Farnborough. Capper was soon to become the leading figure in British aeronautical research, and his main collaborator would be S. F. Cody.

At the beginning of 1904, Cody was still trying to find a sympathetic ear within the military establishment to fund his further kite designs. Lacking government support, his main source of income was still the theater. In January, he staged another new play, *Nevada,* but once again failed to match the popularity of *The Klondyke Nugget.*[1] Though *Nevada* appears to have quietly disappeared, the Codys continued to book engagements for *The Klondyke Nugget.*

Cody moved his workshop from Alexandra Palace to the Crystal Palace grounds, which had become a new locus for aviation activity. The venerable Spencer ballooning firm was located nearby, and Hiram Maxim was often on the grounds, working on an airship amusement ride. One of Cody's last kite launches at Alexandra Palace sent his stepson Leon in a man-lifting kite up to a height of two thousand feet. On separate occasions, Francis A. Barton, Francis's son, and the son of Stanley Spencer also were hoisted aloft.[2]

The admiralty had not yet lost complete interest in kite capabilities. During the spring of 1904 British ships along the coast used Cody kites in tests of flag signaling. These kites were constructed at Cody's new workshop at Crystal Palace. Over the next three years, naval ships employed Cody kites to test wireless transmissions, but lost several kite kits to the winds during those experiments. Even so, the kite orders Cody secured from the admiralty during this period were lower than he had hoped, and fell far short of the amounts that he needed for more ambitious aviation endeavors.

The letters that Cody had sent to Aldershot were received by Sir John French, the commander of Aldershot Military District. French, a veteran of the Boer War, was fully aware of the strategic value of reconnaissance. After meeting Cody for the first time, he also took a personal liking to him. French suggested to Colonel J. E. Capper, the superintendent of the Balloon Factory at Aldershot, that he should ask Cody for photos and descriptions of his kites. Capper, another veteran of the South African campaigns and an experienced balloonist, already knew of Cody and the tests he had done with the admiralty.

Capper arranged for Cody to put his kites through a series of tests at Aldershot during the month of June 1904. The tests were assisted by officers and staff from the Balloon Factory, and concentrated on man-lifting capabilities for observation purposes. These trials proved successful, and Colonel Capper recommended both the purchase of man-lifting kite kits and the hiring of Cody as a consultant.

As the de facto head of British military aviation, Capper was charged with several tasks. First, he had an immediate obligation to develop effective aerial reconnaissance techniques, using balloons and kites. Second, he was charged with researching craft for controlled flight, including both airships and powered heavier-than-air machines. Capper was monitoring a project that the War Office had commissioned before Capper's arrival at Aldershot: the building of a new, full-scale airship at Alexandra Palace by Auguste Gaudron and Dr. Francis A. Barton. Construction on the Barton airship had begun while Cody still made Alexandra Palace his headquarters, but work had dragged on through 1904 and early 1905 due to a horrific hydrogen explosion that nearly killed Dr. Barton.[3]

In July 1905, the long-delayed first public flight of Dr. Barton's airship took place at Alexandra Palace. Although it appeared to have better flight controls than Gaudron's 1898 design, and was able to traverse several miles, it did not match the performance expectations that the War Office had stipulated. The Barton ship was damaged after drifting into a stand of trees, and was not rebuilt.[4] S. F. Cody, along with Hiram Maxim and Colonel Capper, had been on hand to witness the flight. The shortcomings of the Barton craft left Capper with the problem of trying to equal the success that foreigners such as Albert Santos-Dumont, Thomas Baldwin, and Ferdinand von Zeppelin were achieving with their airships.

Capper had been impressed by the work of the Wright brothers when he had visited them in 1904, but had advised against purchasing their machines—he thought the cost the Wrights suggested was too great. Instead, Capper recommended to his superiors that the War Office should support its own heavier-than-air research. As a result, by 1905 Capper had under his supervision several aviation projects involving balloons, kites, airships, and heavier-than-air machines.

Capper's recommendation to hire S. F. Cody as a kiting instructor was finally approved in February 1905, for a limited three-month contract. Cody's acceptance of that position put a halt to his theatrical career. Though he continued to wear his wide-brimmed hat, S. F. Cody would never again take the stage in the role of a frontier character. Stepsons Leon and Vivian joined Cody at the Balloon Factory's new headquarters in Farnborough, near Aldershot, while Edward continued to pursue a career in the theater.

With Cody guiding them, the men of the Balloon Factory put their kites through another battery of tests. They tested first man-lifting, then photographic rigs, then improved winch mechanisms for hauling in the tether cables. Cody built a special kite for flight during storms. Together, they tried night launches. They drilled to see how quickly launches could be made, and how speedily a descent and packing for transport could be accomplished. Cody's consulting contract was renewed for another three months in May, and again in August. He was then given a five-month extension, and finally, in April 1906, a two-year contract.

S. F. Cody's ambitions beyond kites became manifest in July 1905. He built a man-carrying glider with a fifty-one-foot wingspan and separate ailerons for controlled turns. Cody's glider was launched as a tethered kite, but the pilot, lying flat on the lower of the bi-wings, could release the guide wire and glide freely to the ground. The glider was flown first at the Crystal Palace grounds, then at Aldershot. Vivian piloted its last flight, when the glider stalled and crashed. The decision was made not to rebuild it.

Although the War Office had finally given Cody some degree of financial security with his two-year contract, they continued to balk over paying him royalties for leasing his kite patent. The military's legal team contested his patent, and ultimately Cody accepted a renegotiated agreement. With that matter settled, the Codys moved to a house near Aldershot and gave up their workshop at Crystal Palace. S. F. Cody celebrated by purchasing a motor vehicle, a Simms-Welbeck. In 1906, automobiles were still limited to mechanical enthusiasts. Cody drove it away without bothering to receive any operating instructions.

Colonel Capper had committed the men of the Balloon Factory to complete a new airship project, built in their own shed. On the basis of Gaudron's and Dr. Barton's experience, Capper realized that controlling an airship required a more powerful, lightweight engine. He dispatched S. F. Cody to an aeronautical show in Paris in December 1906 for the express purpose of finding an engine, since none that fit the bill appeared to be available from British manufacturers. Cody thought the engine might also serve as the driving power for his next design: a motorized kite glider—in other words, an airplane.

Capper however already had another inventor working on a heavier-than-air machine. His name was John William Dunne, a lieutenant in the Royal Engineers and another veteran of the Boer War. Capper likely felt that Dunne was a man in his own mold—a military engineer committed to technological innovation. Dunne was the product of a proud army heritage, being the son of General Sir John Hart Dunne. Dunne was inspired to study the problem of flight by the writer H. G. Wells, whom Dunne admired for the published predictions Wells had made about human flight. Years earlier, Dunne had entrusted Wells with his aeronautical notes before leaving for the war in South Africa.

Capper's frugality in recommending against doing business with the Wrights, combined with his own ambitions and a sense of patriotism, biased him towards giving support to Dunne's efforts. In sending Cody to Paris to find an engine, Capper did not intend to further Cody's airplane efforts. Instead, Capper needed the engine in order to power the Balloon Factory's new airship. When Cody returned with a French Antoinette motor, Capper recruited Cody to be the airship's chief engineer. Although the position was a challenge to his skills, Cody no doubt felt it was a dubious honor, since it took time and resources away from his airplane project.

The military airship, christened *Nulli Secundus* (Second to None), was ready for its first test in September 1907. Colonel Capper was being pressured by his superiors to demonstrate an airship comparable to the efforts of Zeppelin in Germany and Santos-Dumont in France. The new airship's gasbag was sewn from panels of goldbeaters skin (lightweight, nonporous cattle intestines) and formed into a symmetrical cylinder with conical ends. In comparison to other airships of the time, the outline of the *Nulli Secundus* was elegant. Two maiden flights were made at Farnborough on September 10; after these tests Cody added additional stabilizing aeroplanes for better control.

On October 5, hundreds of thousands of London residents were stunned to see the *Nulli Secundus* flying over the city.[5] It had come dozens of miles from Farnborough, and Cody and Capper steered it over Buckingham Palace, above the War Office building, and around the dome of St. Paul's Cathedral. Crowds filled streets and parks to gawk at the flying machine. Cody and Capper tried to land at Clapham Common, but the throngs of people on the greens there made a landing too risky; instead, they chose to alight at the Crystal Palace grounds.

To the general public the airship's flight was a national triumph. The craft was however left exposed to the elements while tied down at Crystal Palace, and had to be hastily deflated by tearing the fabric after a squall of wind threatened to break its moorings loose. Capper, instead of being lionized, was blamed for taking inadequate care of the ship and for wasting extremely expensive hydrogen. Capper and Cody returned to Farnborough, where Capper was obligated to quickly finish a second model, the *Nulli Secundus II*. Meanwhile, Samuel F. Cody was becoming increasingly irritated that his services were not being rewarded with what he considered sufficient compensation, and he was also upset that the airship venture had distracted him from construction of his airplane.

Maud Lee spent the fall months of 1903 in Chicago's Bridewell Prison, unaware of the legal firestorm her arrest had caused. Frank Butler and Annie Oakley began filing their fifty-five libel suits in November 1903, seeking between $75,000 and $100,000 from each newspaper that had printed the story.[6] The suits shocked the editorial boards of these newspapers, and their initial reaction was to deny any fault for printing an erroneous news service story. They blamed the news service for not checking facts or the originating reporters such as George Pratt for misinformation—or Maud for being the deceitful impostor who started it all.

After the Butlers angrily responded to the published stories, further news accounts appeared that vilified Maud. She was labeled as an "impostor," "badly diseased," a "wretched creature," a "toothless hag," "a mental and physical wreck," "a low down tramp," and an "old, soiled, obviously coked-up cowgirl." One account said that she was charged with an "indescribably sordid misdemeanor." Another paper concluded "it seems incredible that one woman could wish to attach another's name to her own blackened character."[7] Unlike Annie Oakley, however, Maud was not in a position file libel charges.

Annie Oakley's complaints to the courts stated that her good name and credit had been permanently injured by the stories. Moreover, she said that she suffered great mental anguish and bodily pain from the shame that the account unjustly brought down upon her, and that her health had been impaired to the extent that she was physically and mentally incapacitated from pursuing her professional career and earning a livelihood. Over the next six years, many would accuse Annie of seeking to mine the cases for windfall rewards, but those who so accused her could not have known how hard the young Annie Mosey had worked to pull herself and her family out of impoverishment. Annie Oakley did not seek compensation—she sought vindication.

The legal defenses that the newspapers' attorneys presented were varied and complex. A common defense was that the articles did not fit the usual legal definition of libel as publication of an untruth with malicious intent. They denied that any harm against Annie Oakley was intended— they were just selling newspapers. That strategy was usually unsuccessful, since precedents existed that the motive to sensationalize a falsehood for profit was malicious.

In some cases the newspapers attempted to place the blame squarely on Maud, arguing that she provided answers to questions that verified that she was Annie Oakley; they also argued that her Chicago jailers identified her as Annie Oakley. That defense was shaky, too, because Maud was arrested and named in court dockets as Elizabeth Cody (her stepmother's first name) or Lillian Cody, but while in custody never gave her name as Annie Oakley. In these cases the Butlers' lawyers argued that the story Maud told while in jail, including the name she gave, differed in many obvious respects from the career of Annie Oakley, which should have been caught by even cursory fact checking.

Some newspapers mounted a defense that Maud had indeed performed as Annie Oakley (in the burlesques in England); that others called her by that name; that she had been billed under that name; and that Mrs. Butler (Annie) had not established a legal claim to exclusive use of that name. Therefore the news story was accurate in that it related to Maud's "Annie Oakley" character, not Mrs. Annie Butler's. This defense was also flimsy, since the news stories clearly stated that Annie Oakley had been with Buffalo Bill's Wild West, had performed before King Edward, and had performed at the Chicago World's Fair—none of which was true of Maud.

After Maud was released from prison, she was deposed at the courthouse in Norristown at least once, and was called into court as a witness on another occasion. Frank Butler was in the room to listen to her Norristown deposition. In her deposition, Maud gave a rational account of her background, which did much to explain how she could have been misunderstood at the time of her arrest. Her deposition omitted any mention of her being related to Buffalo Bill Cody, or to Vivian being her own child. Again, whether Maud really made statements to that effect when she was arrested depends on one eyewitness—the jailhouse matron. Frank Butler, as he listened to her deposition, was perhaps able to make more sense of her career and identity than any other person.

As far as Maud's court appearance, one has to wonder whether Annie Butler and Maud Lee glanced at each other across the courtroom with any sort of recognition. Maud testified that she had met Annie once before in early 1890, the period when S. F. Cody and Annie were performing in the ill-fated *Deadwood Dick*.[8] The fact that Maud was not compelled to testify more frequently probably reflects a belief by both sides in the case that her account might be used against either party. The last of the cases was settled in early 1910, six years after the first cases were tried. By that date, Maud was no longer available to testify.

After leaving Chicago's Bridewell, Maud returned home to her parents' house in Swedeland. In all probability, they pleaded with her to "quit the business" and give up her performing ambitions. Her guns and other tackle had disappeared when the Forepaugh-Fish Wild West Show's assets were dispersed, so she had none of the tools of her profession.

On December 8, 1904, a small court report appeared in the *Philadelphia Inquirer:* "Mrs. F. S. Cody . . . , who is regarded as one of the greatest pistol shots in the country, was convicted in Criminal Court today of the larceny of coal from cars on the Reading Railway and sent to jail for thirty days." There are several possible reasons Maud could have stolen the coal: for heat, to sell for money, or perhaps in order to mold a new batch of target balls.

After Maud was released from jail in January 1905, she again returned to the home of her father, mother, and aunt. Despairing for her future, her family apparently decided upon a different solution to Maud's reckless behavior. In April 1905, Maud was married to a young, illiterate Italian immigrant named Gaetano Fontanelli who lived in nearby Bridgeport, Pennsylvania.[9] Nothing is known about Fontanelli, not even the correct spelling of his name—"Fontanelli" is the most common variant. Bridgeport had a large immigrant Italian community, with many men taking jobs in the quarries and foundries near Norristown.

Maud admitted a previous marriage on the marriage license application, but indicated she had been widowed in January 1902. The illegality of her second marriage hardly mattered, for within a few months Maud was back living with her parents, and Gaetano Fontanelli was not heard from again. Given that Maud's mental condition had worsened, it is likely that her new husband was overwhelmed by her behavior and turned her back over to her family.

By the early months of 1906, Maud's behavior had become so erratic and potentially dangerous that she was admitted to the Norristown State Hospital, the state-operated mental asylum of eastern Pennsylvania. She was diagnosed with *dementia praecox,* a condition that is today usually recognized as schizophrenia. Maud's family was either driven by desperation to have her committed or saw her forced there by a court order. The Norristown State Hospital, and Pennsylvania's asylum system as a whole, was not a place where patients were sent for effective therapy. In fact, in 1906, Pennsylvania's mental health system was in crisis.

In March of that year, the Women's Department at Norristown had nearly 1,500 patients, 300 more than the intended capacity of the facility. There were not enough beds for all the patients, so some slept in hallways. The kitchen lacked adequate stoves, ovens, pots, plates, utensils, and sinks to feed that number. There were insufficient bath facilities and toilets. In one ward, sixty-eight patients shared just one bathtub, and no showers.[10]

The hospital administrators did not attempt to hide these conditions. If anything, the managers of Pennsylvania's asylums tried to bring the state legislature's attention to the deplorable conditions. Indeed, their biggest problem in providing adequate care was in finding hospital attendants willing to work in the nightmarish conditions. The 1906 Annual Report of the Norristown State Hospital to the state said, "The only hope that is left is the arranging for such relief from overcrowding and such improvements in the facilities for work that the care of the patients may be made less burdensome" (p. 82).

At the same time that Maud was admitted to Norristown State Hospital, newspaper editorial boards around the state were calling for new and larger asylums. The *Wilkes-Barre (Pennsylvania) Record* declared, "Such conditions are worse than a shame—they amount almost to an infamy."[11] The outcry spurred the formation of the State Commission on Investigation of Hospitals. In December 1906, the commission visited the Norristown facility. While they expected overcrowded conditions, they were shocked to find cots placed so close together that nurses could not step between them; they saw patients sleeping on floors and people being herded into rooms lacking any ventilation. Every member of the commission vowed that relief should be made as soon as possible.[12]

The only positive note was that the hospital staff and administrators were never accused of indifference or cruelty; instead, they were praised for putting the concerns of their patients first. They viewed themselves as medical personnel treating people with diseases, not as wardens of inmates. Even so, there were few effective pharmaceutical sedatives, so violent patients had to be physically restrained. By 1906, the use of prolonged hydrotherapeutic baths had come into vogue as a way to calm the agitated, but Norristown lacked enough tubs. A variant technique was to swath the patient, mummy-like, in wet cloth strips.

In February 1906, just a few weeks before Maud was admitted, the hospital suffered an outbreak of scarlet fever. This disease, caused by a type of strep bacteria, is now treated by antibiotics and also appears to have lost virulence; but in 1906 few measures worked to stop its spread. The epidemic added to the Women's Department annual mortality rate of 6 percent in 1906, which was still lower than the rate in previous decades.

About 15 percent of the female population recovered their mental health sufficiently to be released each year. The average length of confinement was between four and five years, but that average collapses large numbers of shorter stays on one end and of longer stays on the other end of the scale. Maud Lee, registered as "Maud M. Fontenelli," was not among those who were released after a short confinement. The largest age group among patients was those between thirty and forty years old. Maud was thirty-three when she was admitted.

Against the horror of overcrowding, it should be noted that the patients were provided with food, dental services, eyeglasses, physicians, and emergency surgical procedures. They had access to a library, including magazines and newspapers. There were scheduled recreational events such as lectures, slide shows, parties, balls, picnics, and band concerts. At Christmas they were given toys, gifts, and chocolates.[13] However, the visiting legislative commission members could not find any patients that wanted to remain there.

CHAPTER 11

SOLO FLIGHTS

1908-1913

By early 1908, Cody's relations with Colonel Capper had been strained by haggling over the amount of money the War Department owed to Cody for the kites he had supplied. Cody had asked for £5,000 plus expenses; Capper had recommended £3,000 plus another £1,100 for expenses. The War Office initially recommended that Cody should receive only between £500 and £1,000, but the Ordnance Office agreed to pay him £1,000. However, the Treasury knocked the figure back to £500 in August 1908. Cody was understandably indignant.

He was also frustrated at the slow progress being made on his airplane; he was still waiting for an engine. The Antoinette motor from the first *Nulli Secundus* was designated for use in the Balloon Factory's next airship, the *Nulli Secundus II*. Not until July 1908 was a second Antoinette motor acquired.

Cody spent the month of August conducting more kite trials aboard admiralty ships, but once again news from the Wrights changed the dynamic of aviation efforts around the world. Wilbur Wright had brought the Wright Model A to Paris to demonstrate its capabilities to a skeptical European aeronautical community. Little had been heard from the Wrights in the previous five years, and many assumed that they had not progressed much beyond their Kitty Hawk achievement—which was still suspected to have used assisted takeoffs and which lacked both duration and control.

Alberto Santos-Dumont had been able to match the Kitty Hawk effort in the fall of 1906, as had Frenchmen Leon Delagrange in 1907 and Henry Farman in early 1908. In North America, the Aerial Experiment

Association founded by Alexander Graham Bell developed three models (*Red Wing, White Wing,* and *June Bug*) that had flown short flights by mid-1908. Therefore, when Wilbur Wright arrived in Paris, observers were not expecting to be greatly impressed.

Wilbur's series of Paris flights showed however, that the Wrights were far ahead of any other airplane designers in the world. Wilbur was able to turn, fly circles and figure eights, fly at a higher altitude, and fly much longer than anyone else had yet done. A few weeks later, Orville Wright made even more impressive flights for the U.S. Army at Fort Meyer, near Washington, D.C. The Wrights' masterly displays sent shock waves through the aviation world—and alerted the British government that it was lagging behind in this strategic technology.

The success of the Wrights' public flights was doubly embarrassing for Colonel Capper, who had recommended against licensing their designs. Capper's faith in John W. Dunne's airplane project was not rewarded. In late 1907, Capper and Dunne conducted secret tests at Blair Atholl in Scotland, the estate of the Duke of Atholl. Dunne's ambitions were hampered by an underpowered fifteen-horsepower motor and by his own poor health. Dunne's airplane had swept-back wings, which he believed to have a great advantage in providing stability. In late 1907, however, his airplane crashed as it was executing an assisted takeoff while sliding down a mountainside ramp at Blair Atholl. In later tests in 1908, the Dunne plane made several short jumps but was obviously a huge disappointment compared to the impressive accomplishments of the Americans and the French.

Therefore, a somewhat desperate Colonel Capper allowed Cody to proceed with his airplane project, despite their chilled relationship. Cody finally was able to install the motor he needed in September 1908. The new craft was designated British Army Aeroplane No. 1. Test runs along the ground at Farnborough were made in the closing weeks of September to tune the machine's balance. Cody made a short jump off the ground for about two hundred feet on September 29, but concluded that he needed a larger propeller.

On October 14, Cody made two more jumps of about three hundred feet each, both too short to be considered official. Then, two days later, on Friday, October 16, 1908, Samuel F. Cody nosed his airplane thirty feet above the ground and flew a quarter mile. With this flight, Cody became the first person to fly an airplane in Great Britain. Unfortunately, on landing, the left wing dipped to the ground and the craft spun around, heavily damaging both wings.[1]

Cody's flight was not noted with any fanfare; his achievement paled compared to what the Wrights had done in August. The wreck forced Capper and Cody to rebuild a completely new airplane, which consumed the remaining weeks of 1908. British Army Aeroplane No. 1B was ready by early January 1909. During its test runs along the ground, hops of a few dozen yards were made, but the positioning of the radiator and support surfaces needed realignment. During these tests, Colonel Capper warned newspaper reporters that only small hops would be made, but that did not stop them from reporting that the new craft was a failure. Their assessment seemed accurate after Cody smacked the airplane to the ground following a jump where he reached an altitude of twenty feet.[2]

Although the jumps and dips looked haphazard to the curious public at Farnborough Common and Laffan's Plain, Cody was methodically testing the controls and learning how to make safe landings. By late February, he was making straight flights of nearly six hundred yards, but was still not confident about attempting a turn.

On February 24, 1909, both S. F. Cody and J. W. Dunne received letters terminating their services at the end of March. The War Office had decided that their efforts were not worth the expense, and that private companies (domestic or foreign) would provide the military with future airplane designs. Cody wrote a letter of appeal, but must have known there was little chance the decision would be reversed. He further pleaded to be allowed to keep the airplane in order to continue testing it as a civilian.

The War Office agreed to give Cody the airplane, but the craft's engine was still the property of the government, so Cody could use it only on a lease basis. He was also allowed to erect a shed at Laffan's Plain, the flat parade ground near Farnborough. The airplane was renamed the Cody 1. Capper made it clear that Cody's assistants were not to mingle with any workmen from the Balloon Factory. Working independently, Cody made longer flights throughout April and May. On May 13, the Prince of Wales visited Aldershot to observe military maneuvers, and afterwards made a side trip to nearby Farnborough to see Cody make a short flight.[3]

By June, Cody was able to successfully execute turns. On June 18 he flew more than two miles in a circular path, at an altitude of twelve feet and a speed of twenty-five miles per hour. Reporters who were observing again mistook some erratic motions during his flights as proof that the airplane was underpowered, and privately Cody did confess the need for a stronger engine; but that was not the reason his flights looked wobbly—he was still systematically checking the controls.[4] A month later, he was able to fly a total of four miles and execute three complete circles.

In August Cody rebuilt the airplane using a new French ENV engine, which provided about twenty more horsepower than the Antoinette engine. In the redesigned configuration, Cody sat in the forward part of the plane, his vision unimpeded. "I find my new position in front of the engine has a much more sensational effect on the nerves than the old position," he noted. "In fact, until last night I never knew that I had any nerves. I think, however, I shall get over this slight timidity after a few runs."[5]

With the added power, he was able carry a passenger behind his seat. He gave the first honors to Colonel Capper. Then he took Lela for a pass over the field, making her the first female airplane passenger in Great Britain.[6] After landing, Cody turned around in his seat and kissed her. In the last weeks of August, he in turn took up his son Franklin and his adopted sons Vivian and Edward. By the end of August, he was making flights of eight miles, lasting an hour and a half—matching the feats accomplished by Wilbur Wright in Paris a year earlier. On September 8, he set a distance record of more than forty miles.[7]

Without the support of the Balloon Factory and with the end of his contract as a kiting instructor, however, Cody had no source of income. The dispute over compensation for the army's use of his patent still simmered. Cody was left with the necessity of somehow earning money for his aviation efforts. The most immediate prospect in that regard was to enter prize competitions and air meets. Such competitions had proved popular in Europe. Newspapers found that these races boosted their readership, so they were willing to offer lucrative prizes. Some of the competitions were limited to British pilots or British airplanes, however; as an American, Cody could not qualify.

In the fall of 1909, Samuel F. Cody rectified this problem by taking steps to become naturalized as a British citizen. It must have been a difficult decision, given that his entire public persona had been based on being a brash American cowboy. On the other hand, he had been away from the United States for nearly twenty years and he now had a British family. On his naturalization application, Cody repeated his usual deceits about his birth year (which he gave as 1862), birthplace (given as Birdville, Texas), father's name (Cody), and marital status. He was however honest about the reason he wanted to change nationality—he said that he did so in order to further pursue his career.

With only one viable airplane, Cody risked not only his life but his livelihood each time he made a flight. On some occasions he landed roughly, with minor damage to both himself and his machine. He deemed some of the offered prize competitions too risky, turning down invitations to fly across the English Channel and to race in France. He flirted with vying for a prize offered for a London-to-Manchester flight, but decided that it was liable to be beyond the limits of his current design.

He did accept a spot in the Doncaster Air Meet, in which he and several other flyers were promised appearance fees. The other pilots were Roger Sommer, Hubert Le Blon, Leon Delagrange, Walter George Windham, and J. G. Lovelace—all of whom flew French-inspired designs of small, lightweight monoplanes. Cody's biplane was a behemoth by comparison. The weather during the meet was wet and windy, which kept the pilots on the ground and the spectators anxious for activity. Cody felt obliged to try a flight, but on landing he hit a soft spot in the ground, causing his machine to pivot onto its nose. Repairs were made, and Cody was able to make a few more short flights over the remaining days of the meet. To compensate for the underwhelming spectacle, Cody made a great show of signing his naturalization papers before a crowd, with a brass band providing accompaniment.[8]

In December 1909, Cody attempted to win a £1,000 prize for making a flight from Liverpool to Manchester. Due to fog, however, he had to abandon his attempt after covering just thirteen miles. His appearance money from Doncaster allowed him to begin work on a new model in January 1910. This design, in order to meet the "all-British" require-ments of some competitions, would be powered by a motor manufac-tured in England by August Green. The new machine took nearly five months to complete—precious weeks at a time when there was a furious pace of development in aeronautical engineering. Cody's status as a vir-tual one-man operation was beginning to look like an anachronism.

In June 1910, during the Cody Flyer IIA's shakedown tests, Cody nosed the new machine into the ground from a height of about forty feet. Initial reports claimed that he was mortally injured, but he made a rapid recovery and was flying again in July.[9] He made steady progress in making longer flights, all the while modifying the machine with differ-ent engine configurations. In the fall of 1910, the Cody Flyer IIC estab-lished a leading distance in the all-British Michelin Cup competition of 94½ miles. This mark was quickly bested by two other fliers, however, and all three hoped to set new marks on the very last qualifying day, December 31. Cody emerged the victor with a flight of 4 hours, 47 min-utes that covered 185.46 miles.

By early 1911, Cody's dramatic aerial successes had endeared him to the British public. This affection was confirmed in January when a waxwork effigy of Samuel F. Cody was commissioned for London's oldest and most famous wax museum, Madame Tussauds. When Cody and his first wife, Maud Lee, had started their careers, they played in American dime museums that featured waxworks, so in one sense the installment of his likeness at Madame Tussauds would be a return to his show-business roots.

The prize winnings that Cody had garnered allowed him to complete a new, third airplane in the spring of 1911, the Cody Mark III. In the eve-ning of June 6, Cody demonstrated the new machine for King George V at Laffan's Plain. On descending, Cody flew directly over the royal party at an altitude of twenty feet. He was later taken to task in the newspapers for endangering the life of the British monarch.

Cody was preparing the Cody Mark III for the *Daily Mail*'s Circuit of Great Britain Race, one of the longest and most lucrative multistage prize races of its era. The first stage started in Brooklands, south of London. From there the route headed five hundred miles northward through eastern England and into Scotland, to the northern terminus at Stirling. There, the racers would turn southward and traverse western England, all the way to Exeter in the southwest. The final leg would follow the southern coast before a last turn north brought them back to the starting point. The total distance was 1,010 miles; the prize was £10,000 pounds. For Cody, the challenge was irresistible.

Thirty pilots entered the race, twenty-one showed up at the starting line, and two of them failed to take off. Cody's Mark III was smaller than his first two machines, but still monstrous compared to the vehicles flown by his competitors. He also was still using his Green sixty- to eighty-horsepower motors, which meant his machine was slower than the other entries. Two of his opponents, Jules Vedrines and Jean Conneau, were experienced champions of Paris-Madrid and Paris-Rome air races. The only advantages Cody had were an intimate knowledge of the mechanics of his machine and the goodwill of the English people.

The flyers set off on July 22, with several members of European royalty watching at the starting line. It took Cody six days to reach the northern terminus at Stirling. Two of those six days were spent in forced idleness due to damage he accrued taking off from and landing on soft fields. At that point, less than halfway through the course, Cody had already lost the race. Conneau had completed the circuit in just five days, with Vedrines just a few hours behind him. Cody was informed by race officials that he was one of eight remaining competitors: two others were still ahead of him, and five were behind. The five flyers behind him quickly retired from the race.

Although the grand prize was beyond his reach, there were many good reasons for Cody to continue. Foremost among them was that he was generating enormous publicity for himself, which was both enjoyable and could be useful to him in the future. The Circuit of Great Britain Race also had small incentive prizes that Cody could still earn or share, such as being the first British pilot in a British machine to cross the finish line. When he reached the checkpoint at the town of Filton, he shared an award for having his original motor intact. Finally, it would have been within his character to have placed side bets on himself to provide an extra incentive.

Cody completed the circuit on August 5, the last day that a finish would be recognized. He came in fourth overall—a better showing than anyone had predicted for him. The attention he had gained solidified his reputation as Britain's foremost aviator, and his perseverance made him even more beloved to the general public.[10]

In the fall of 1911, Cody vied for two trophies offered by Michelin, after having won the Michelin Series 1 prize the previous year. Michelin added a separate trophy in 1911 for the pilot who could cover a distance of 125 miles in the quickest time. To Cody's own surprise, he won that prize on September 11, 1911, with a time of 3 hours, 6½ minutes in his Cody Mark III. Then, on October 29, Cody won the original Michelin trophy for the second year in a row, for completing the longest sustained distance flight: he covered 261½ miles.

Despite Cody's fame, no buyers approached him to acquire any of his designs, whereas many people were purchasing Bleriots, Farmans, and Wrights. In January 1912, in a move that sounds desperate, Cody wrote to the government of Australia, offering his services as an airplane and airship builder and instructor. He received no positive response.[11]

In February, Cody's seven-year battle over compensation owed for supplying the army with his kite designs was finally brought before Chancellor of the Exchequer David Lloyd George. While his case was being arbitrated over the first six months of 1912, Cody earned small amounts of income by giving flying instruction to students and by giving exhibitions. He also followed through with his patent applications, perhaps hoping that they might generate royalties.

Cody realized these small and infrequent payments would not be enough to keep him in the game. Therefore, he pinned all his hopes on an event that he entered with a distinct disadvantage: a competition to decide the type of craft that the Royal Flying Corps (RFC) would adopt when it was scheduled to form in mid-1912. The War Office bureaucracy with which Cody had waged battles over the past seven years set rigorous standards that the winner would have to meet, with an emphasis on safety and efficiency. The Military Aeroplane Trials were scheduled for August 1912.

Cody spent the first half of 1912 cobbling together an income, but he was also busy building a new airplane specifically for the Military Aeroplane Trials—a sleek monoplane that would be designated the Cody IV. He had the good luck to acquire a powerful 120-horsepower Austro-Daimler engine from another pilot, which he hoped would finally dispel his reputation for underpowered machines. The new machine was completed in June 1912, and Cody moved quickly to fine-tune its performance over the next month.

Unfortunately, Cody experienced a blow in the first week of July when one of his flight school pupils, Lieutenant H. D. Harvey-Kelly, crashed the Cody III biplane. This left Cody with only one craft at his disposal—the new Cody IV monoplane. Cody had hoped to use the Cody III as a second entry in the Military Aeroplane Trials, since it had the capacity to carry passengers, and Cody expected that one of the trials might require this feature.

Just a few days later, Cody was flying the monoplane when his engine cut out at two thousand feet. He glided in for a landing at Laffan's Plain, but as he hit the ground and rolled along, a cow that had been startled by the sound of the craft ran directly in front of the propeller. Both the cow and the monoplane suffered mortal wounds, although Cody was only dazed. Thus, he was left with no working airplane and the Military Aeroplane Trials due to start on August 1.

What followed in the remaining weeks of July was one of Cody's defining moments. He gathered his sons and helpers, telling them that he intended to salvage components from the two ruined machines and create a new biplane, the Cody V. By necessity it would be a new design—and they had three weeks to get it from the drawing table to the field. The one asset in his favor was the Austro-Daimler engine, which had survived the run-in with the cow. Cody and his crew worked round the clock and, miraculously, had the craft ready by July 31. It was a superhuman feat that still astonishes aviation historians.

As if to signal that his luck was turning—and Cody was nothing if not superstitious—on August 1 he received word that the Lord Commissioners of the Treasury had decided the arbitration proceedings in his favor, awarding him £4,000 in payment for his kite designs and supplies. The money came too late to assist with the Military Aeroplane Trials, but it put an end to his long fight for recognition for the assistance he had given.

Thirty-one different airplanes were registered to participate in the trials, but on the starting date only nineteen were able to fly. Entries were permitted from foreign as well as from British designers, and included the most famous manufacturers of the era: Bleriot, Farman, Bristol, Vickers, and Handley Page. An award of £4,000 was allocated to the overall winner, and an additional £1,000 set aside for the best British entry. The judges' requirements were as follows:

1. The ability to carry a passenger and offer an unobstructed forward view
2. An easily started motor
3. Dual controls, for training of pilots
4. Shelter from the elements for pilot and passenger, and the ability of the two to communicate via voice
5. Interchangeability of all parts to allow for repairs
6. Ability to pilot the machine without exceptional strength or strain
7. Capability to quickly dismantle, rebuild, and crate the machine for easy transportation.

Additionally, the airplanes would be tested for speed, climbing ability, top altitude, gliding, gas and oil consumption, and landing distance.

Many of the flying tests that took place in August gave an advantage to Cody's state-of-the-art Austro-Daimler engine, which made up for other faults in his machine's hasty design. At the conclusion of the trials, Cody's entry had the winning score in both the overall competition and the British-only category. Cody won not only £5,000 in awards, but also a guaranteed order of two airplanes for use by the new RFC.[12]

Cody spent the remainder of 1912 basking in his critical and financial successes. In September, his waxwork figure was unveiled at Madame Tussauds. It depicted Cody in his aviator clothes, with his clear eyes staring intently up towards the sky—or the future. Many years later, some would say it looked like Vladimir Lenin in an idealized propaganda pose.

Cody's year was capped with another win of a Michelin prize: he had the fastest time in covering a set distance of 186 miles.

As he entered 1913, Samuel F. Cody finally had enough capital to take steps to establish a publicly held company, Cody and Sons, Aerial Navigation. Fifty thousand shares were put on sale at £1 apiece. Using this infusion of investment, Cody hoped to build a dozen more factory sheds at Farnborough. Moreover, he began working on a new aircraft, the Cody VI, which could be adapted with floats to be a seaplane.[13] Cody made public his interest in vying for the recently announced £10,000 prize for a flight across the Atlantic—although that would require a design beyond the Cody VI.

The Cody VI was built specifically for Lord Northcliffe's £5,000 Coastal Circuit of Great Britain Race, scheduled to begin on August 15, 1913. Cody's airplanes were noted for their large size, but the Cody VI was his largest biplane yet, with a wingspan of fifty-nine feet and a length of more than forty feet. Cody began test flights with normal wheeled landing gear in the summer of 1913. In early August, during the bank holiday, Cody flew to Brooklands to give a public demonstration of his new craft. He and Leon then flew it back to Farnborough.

At Farnborough, Cody finalized his plans to refit the Cody VI with floatation gear in preparation for the Coast Race. On August 6 he gave a ride to August Green, whose company made the engine installed in the Cody VI. On the morning of August 7, 1913, Cody took on Lieutenant Keyser as a passenger; on landing, he agreed to give another ride to Keyser's friend, the famous cricketer W. H. B. Evans. They flew for eight minutes over the local golf course, then turned back for a landing. At an altitude of about two hundred feet, passing over Ball Hill, the craft seemed to collapse along the center of its length. Neither Cody nor Evans had seatbelts, and both were thrown out of their seats. They fell to the ground, dying instantly.[14]

A coroner's inquest was conducted at Aldershot two days after Cody's death. Witnesses to the accident gave conflicting accounts of the sequence of the machine's disintegration, and there were questions about whether pieces of the crash debris might have been moved or stolen by souvenir collectors before investigators made their examination. Cody's stepson Leon was convinced the propeller had snapped and torn through the wing fabric, setting off a catastrophic failure, but other witnesses thought they saw the rear section tear away first. The inquest was suspended for two months pending further investigation by experts.[15]

General Sir Douglas Haig, the commander at Aldershot, took a leading role in the funeral arrangements. Though Cody was no longer attached to the military in any capacity, he was granted a plot in the service cemetery at Thorn Hill, the first civilian so honored. On the afternoon of August 11, 1913, Cody's coffin, covered by the Union Jack, was conveyed by a carriage drawn by six black horses from his home in the village of Ash Vale to the cemetery chapel a few miles away. More than a thousand soldiers from Aldershot and the RFC at Farnborough volunteered to form a procession. In all, an estimated fifty to one hundred thousand people lined the road along which Cody was carried. One hundred large floral tributes were conveyed alongside Cody, with the procession being led by kilted members of the Black Watch.[16]

The newspapers had run articles in Cody's memory in the days between his death and burial. The news had quickly reached America, but was not accorded any more attention than other aviator fatalities. No country but Great Britain could claim Cody as a hero. King George V issued the following statement: "I have received with profound regret the news of the death of Mr. Cody. I saw him on several occasions at Aldershot and always appreciated his dogged determination and dauntless courage. His loss will be much felt at Aldershot where he did so much for military aviation."[17]

CHAPTER 12

LEGACIES

1913–1947

Immediately following Samuel F. Cody's fatal accident, his stepson Leon publicly stated that the family had been left in dire financial straits. True, Cody's death meant the end of his fledgling business and the need to auction his sheds and tools. However, Cody had invested much of his Military Aeroplane Trial award, his military kite reimbursement, and Michelin prize money in various stock shares, so Lela and her grown children (Frank, the youngest, was now eighteen) were hardly destitute. There was however strong popular sentiment that Cody had long suffered ill treatment and had never been properly recompensed, which caused many to rally behind Leon's pleas and raise collections for the family.

In America, the father and stepmother of Maud Lee reacted to the news of Cody's death by immediately letting the Camden and Philadelphia newspapers know that they would fight on Maud's behalf for Cody's estate.[1] They suspected that Cody had left some wealth behind. Over the next months, the Lees took steps to secure legal assistance in pursuing this claim on behalf of their daughter, who remained incapacitated.

London's theatrical community was particularly generous. On September 13, 1913, a matinee at the London Hippodrome featuring forty-four different acts and starring some of the most famous performers of the American and British stage was held to benefit Lela and her children.[2] The afternoon began with a lighthearted overture of Kerry Mills's "At a Georgia Camp Meeting," a syncopated American cakewalk dance often played by Cody's old acquaintance John Philip Sousa.[3] Also performing before the audience was Chung Ling Soo, the marvelous Chinese magician, who in reality was the American conjuror William Ellsworth

Robinson. Robinson was an imitator of a notable Chinese magician, Ching Ling Foo, and even used many of his tricks. Though derivative, Robinson's act became the more popular of the two. Robinson died onstage in 1918 while performing his signature trick, catching a bullet fired directly at him. His prop gun was not supposed to load the bullet into the chamber, but on that occasion the mechanism failed.

The Shakespearean actress Evelyn D'Alroy was farther down on the program. D'Alroy would herself die at an early age, thirty-three, just two years after the Cody benefit. Two musical comedy stars, G. P. Huntley and Harry Grattan, then gave a nod to Cody's shooting act by doing a sketch set in a gunsmith's shop. Later during the matinee, the brooding, dashing actor Lou Tellegen and two colleagues produced act 2 of Oscar Wilde's *Picture of Dorian Gray.* Tellegen was the young lover of the aging beauty Sarah Bernhardt. Bernhardt herself was initially announced as a performer in Cody's benefit show, but her name did not appear on the handbill. The age difference eventually split Tellegen and Bernhardt; he went on to four turbulent marriages. After being disfigured in a fire, he was turned down for further acting roles, became depressed, and committed suicide in 1934.

The stage was then cleared for the one hundred singers of the Scala Chorus, the most famous operatic chorus in the world. They sang three selections from Verdi, one from Wagner, and one from Gounod. In a jarring change of pace, these classic selections were followed by Scottish comedian Harry Tate. Tate did a sketch on motoring, in honor of Cody's fondness for his automobile. Tate's life had several intersections with aviation following the benefit for Cody. Using rhyming slang, airmen nicknamed the Royal Aircraft Factory R.E.8 biplane the "Harry Tate." Tate would die in 1940 during the London Blitz. During the show, Leon Cody accepted an Aerial League award on his stepfather's behalf; and two airplane passenger flights were raffled.

The benefit matinee was a monument of the show-business community's affection for Cody. The length of the show and number of famous acts were seemingly out of proportion to the accomplishments of a man whose formal stage experience was limited to one successful production (fifteen years earlier) of a raucous western melodrama. British theatrical society must have adored Cody for his larger-than-life character—off-stage as well as on. Such an outpouring of support would seem unlikely had Cody had a reputation as a coldhearted, mean-spirited person.

Later in September, the coroner's inquest reached a verdict, declaring that the airplane collapsed due to inherent structural weakness. The report also stated that the fuselage section had fallen to the ground intact, and that seatbelts might have saved the lives of the two men. The report damned both the quality of Cody's design skills and his judgment, and seemed to have been written with the intent of discouraging further aeronautical exploits by "amateurs."

Another blow to Cody's legacy came in November 1913, just a few months after his death, when his old friend from the Alexandra Palace, balloonist Auguste Gaudron, died after a long illness. Gaudron was unable to comment on Cody's death and left no memoirs.[4] Had he lived, Gaudron would have been able to clarify Cody's role as Britain's pioneer aviator, as he had earlier defended Dr. Barton's airship innovations. Gaudron very likely might have been able to shed light on how and when Cody started to experiment with Hargrave-type kites. As it stands, Gaudron himself has been overlooked as an early experimenter with airships, a record-setting balloonist, and a successful parachutist.

When Samuel F. Cody died, he had an accurate vision of the future of aviation: he was a believer in large craft with powerful engines; he was seriously considering long-distance, multiengine machines capable of crossing oceans; and he had offered his public opinion that airplanes could be used by the military as tactical weapons as well as for observation.[5] This belief was proven true just a couple of years after his death, when World War I spurred technological leaps in aeronautics. Cody's contemporaries in aircraft design—those with the support of investors and with factory capability—were called upon to define a new age of British aviation: Thomas Sopwith, Frederick Handley Page, Geoffrey de Havilland, and others. Even had he survived, it is questionable whether Cody would have been among their number, as he almost certainly lacked the temperament to delegate responsibilities for running a large-scale manufacturing operation.

Among the RFC enlistees in World War I was Franklin Leslie Cody, S. F. Cody's only natural child. Young Frank Cody faced a devastating mortality rate for airmen. During the first years of the war, German aircraft proved to be much superior to their British counterparts. In January 1917 Frank Cody was flying a mission and found himself surrounded by German fighters. He had little chance to fight and was shot down to his death. His name was just one among thousands in the casualty lists; the fact that he was Samuel F. Cody's son was barely noticed by a populace overwhelmed by loss.[6]

Enno Meyer, the Cincinnati photographer who had captured invaluable images of the Cree and Sioux visitors to his city in 1895–96, continued his interest in using photography to preserve endangered ways of life. Many years after the "Historical Cincinnati" exhibition, Meyer once again felt compelled to visit the Cincinnati Zoo to save a piece of American heritage on his film plates. His subject this time was Martha, the last passenger pigeon on earth. Martha had been the sole remaining example of her species since 1910. On Sept 14, 1914, Martha's feet loosened from her perch and she fell to the ground, dead. *Ectopistes migratorius,* whose dwindling numbers had forced shooting competitions to use glass target balls (and consequently created celebrity sharpshooters), was extinct. Enno Meyer took the last photo of Martha before she died.

The fate of the American bison, the star attraction of Maud's 1902 Great Buffalo and Wild West Shows United, turned out slightly better. Pablo and Allard continued to maintain their Montana herd and offered it to the U.S. government. Although the government declined to purchase the herd, the ranchers were able to negotiate a sale of a large number to the Canadian government. In South Dakota, Scotty Philip cultivated a herd that numbered about 1,200 by the time of his death in 1911. Today, about 350,000 bison survive, but the vast majority are not genetically pure American bison. They are "beefalo": bison crossbred with cattle. A wild herd of about 3,500 roams sections of Yellowstone National Park, and other pureblood herds exist in Utah's Antelope Island and South Dakota's Wind Cave National Park.

The Sicangu Lakota (Rosebud Sioux) hope to reintroduce bison to their reservation in lower South Dakota. Their tribal neighbors, the Oglala Lakota of the Pine Ridge Reservation, are already working toward the same goal. Today there are about 21,000 Sicangu Lakota living on the Rosebud Reservation. The Lakota reservations suffer from some of the highest poverty levels of any community in America. With poverty come high rates of suicide, substance abuse, school dropouts, crime, and gang membership. The remote location of the Lakota reservations precludes the possibility of substantial casino income. Another issue, the restoration of the sacred Black Hills region, is still being contested. The survival of the Lakota's traditional culture remains threatened. An alarming number of tribal members are not fluent in the Lakota language, and thousands of Lakota children are being adopted into non-Lakota families.

There are no more touring Wild West shows to provide an outlet for the survival of Plains Indian culture. Buffalo Bill Cody's show went bankrupt in 1913; it had been on shaky financial ground for many years. Many factors contributed to the demise of the Wild West show: its format became stale, diluted by cheap imitations and the sordid reputation of touring tent shows. Rodeos gained in popularity and were cheaper to produce than theatrical Wild West shows. Movies, which quickly adapted frontier themes into westerns, allowed more flexibility in staging narratives than did the short arena sketches of the Wild West shows.

The most notable twentieth-century successor to Buffalo Bill's Wild West, the Miller Brothers 101 Ranch Wild West, faltered during World War I and survived in a limited form through the 1920s. Small-time shows continued to tour after World War I, and a few Wild West stars found employment as featured performers of circus shows or as rodeo and fair entertainers. Even these vestiges had disappeared by the time World War II started, however.

William F. "Buffalo Bill" Cody died in January 1917. Buffalo Bill had been the celebrity figure whose shadow fell over nearly every aspect of American popular culture from the late 1870s to the late 1890s. For all his public success, Cody suffered from the heartbreaking loss of a child, a rocky marriage exposed by ugly divorce proceedings, and foolhardy business investments that lost him the several fortunes he made from his show. He was however beloved both by the public and within the show-business industry, and he treated his employees fairly and with respect. Toward the end of his career, when he was forced to work as a performer for a show run by circus owners, he was not treated with dignity he deserved.

Adam Bogardus, the first of the celebrity sharpshooters, grew tired of show business and stopped touring in 1891. The loss of his nineteen-year-old son Eugene, who was also touring with a circus, seemed to affect Bogardus profoundly. Bogardus opened shooting galleries in Illinois and Arkansas, gave private shooting lessons, and wrote a book on wing shooting. He still hunted recreationally until shortly before his death at age eighty in March 1913.

Doc Carver, who surpassed Bogardus's feats as a shooter and was Buffalo Bill's first partner, remained in show business until his death at age eighty-seven in 1927. However, he gave up Wild West shows in favor of producing horse diving exhibitions. By Carver's own account, he got the idea in the 1880s after crossing a bridge that collapsed beneath him, causing his horse to plunge into the waters below. Carver did not perform the horse diving act in public until a play written for him called *The Scout* opened in Melbourne, Australia, in 1891. *The Scout* included a collapsing bridge scene that caused a sensation (and likely inspired S. F. Cody to include a similar scene in *The Klondyke Nugget*). *The Scout* opened just two years after Carver had seen audiences go wild night after night over Eclipse, the aerial platform-jumping pony of the Forepaugh Circus, so it may be that Eclipse inspired Carver more than any personal experience did.

Although *The Scout* had a good run both abroad and in America, by the mid-1890s Carver had dispensed with the awkward theatrical trappings and was touring his outdoor diving show, featuring both horses and humans. It became a family enterprise that crisscrossed the country for two decades, then settled in a permanent home at the Steel Pier entertainment center in Atlantic City, New Jersey. Carver died shortly after this move, but the horse diving act remained under the management of his son Al, whose wife, Sonora, was one of the diving horse riders. The horse diving act remained at the Steel Pier until the late 1970s.

Jennie Franklin, aka Jennie Fowler/Moore/Kennedy, stayed only a short time with the Adam Forepaugh Circus in 1888. By 1894 she had dumped husband William B. Kennedy for a new partner, Edward W. Coleman. They toured with an American circus through South America in 1896, where Mexis made worldwide headlines for shooting a target ball from the head of President Joaquín Crespo of Venezuela. Coleman and Mexis continued to perform in small vaudeville venues until 1915, when all mentions of that act suddenly stopped.[7] Jennie, who had been shooting in public for nearly forty years, must have been at least in her sixties at that point. Her fate is not known.

Vic Cody, Maud's protégé who got his professional name thanks to some surplus posters Maud had left over from her partnership with S. F. Cody, continued to perform with small Wild West shows, rodeos, and county fairs. He spent many years with the Kemp Brothers Wild West Show following his stint as Buckskin Bill. In the 1920s he was with the King Brothers Wild West Rodeo. Through all those years his act consisted of pinning a female assistant to a board with his twelve-inch throwing knives and doing shooting tricks. He then joined the Barnett Brothers Circus, where he had charge of their Wild West feature. The last billing found for his act is from 1933.

Annie Oakley and Frank Butler pressed their fifty-five libel suits across the country for six years, from 1904 to 1910. Frank involved himself deeply in the pursuit of these cases; he was present during Maud Lee's deposition by the plaintiff and defendant attorneys at the Norristown Courthouse in late 1905. Frank Butler was probably able to sift more of the truth out of Maud's statements than any lawyer, reporter, or member of the public could have done. Frank and Annie must soon have realized to whom Maud had been married and the course of her career. They came to pity her more than blame her. In 1910, after the last case was settled, Frank wrote

> With the settlement a few days since of two small cases ends Annie Oakley's six years of strenuous litigation in courts of nearly every State in the Union, all caused by a poor wreck of a woman,

a morphine fiend, who when arrested and asked her name, replied that she was the champion shot of the world.

A reporter from a sensational paper and a press association added the name "Annie Oakley," and sent it broadcast, so you see the very word champion, which she [Annie] did not allow managers to use in connection with her name, caused her six years of lawsuits and many days and nights of worry and trouble. . . . I want to say right here that when she entered these suits, it was not money but vindication she was after.[8]

During those years of litigation, Annie performed infrequently, since court dates made it impossible to maintain the regular schedule of a touring show. Additionally, she had injured her back, requiring a succession of spinal surgeries. In 1910, Buffalo Bill invited her and Frank to rejoin his Wild West show, but she declined. In 1911, she did join a rival show, the Young Buffalo Wild West and spent two years with it, before ultimately retiring from arena performances in 1913.

Annie and Frank then moved to the Eastern Shore town of Cambridge, Maryland—a bird hunters' paradise. In 1915 they bought another property a bit farther south in Pinehurst, North Carolina, where they spent their winters. Both Annie and Frank wrote articles on shooting and acted as consultants to firearm industries. Annie gave generously to charities, particularly orphanages, and sometimes performed exhibitions for fundraisers. When the United States entered World War I, she offered to raise a regiment of women volunteers. Her offer was declined, but she did raise money for the war effort.

In 1922, Annie was seriously injured in a car accident. Her recuperation took a long time, and suffered a setback in 1925. Annie, perhaps sensing that she would not improve, moved with Frank back to her hometown in Darke County, Ohio. She died there on November 3, 1926. Frank Butler was devastated by the loss of Annie, and died just eighteen days later. They had been inseparable for forty-five years.

Maud Lee was still a patient at the Norristown State Hospital when her mother, Phoebe Lee, died in 1910. The *Philadelphia Inquirer* posted

Phoebe's obituary notice with the curious statement that Joseph Lee was her brother, not her husband. The next year, Joseph Lee and his wife's sister, Elizabeth Lillian Kay, moved from Swedeland, Pennsylvania, to Camden, New Jersey, where they lived as husband and wife. It is not known if Maud was granted a parole to see her mother's burial, but such furloughs were often granted.

When Samuel F. Cody died in 1913, Maud's father and stepmother/aunt told newspapers that she was not mentally competent to understand the news. Maud might have already succumbed to her own delusion that Cody had drowned while trying to reach the North Pole—a statement she made in her otherwise rational court deposition in 1905. Maud parents had to initiate a court hearing in order to have Maud ruled as being "of unsound mind." With her incompetence established, the way was cleared for the Lees to sue Cody's estate. In May 1914, his estate was valued at £4,259.

In February 1917, a British court finally issued a decision in the Cody estate case, awarding £1,000 to Maud and £250 each to Cody's four Cowdery siblings: a brother and three sisters. What Lela thought of this decision is not known; she made virtually no public statements from Cody's death in 1913 until her own demise in February 1939 at age eighty-seven. Maud's £1,000 was tied up in further litigation for another two years by legal claims against the Lees.

By the time the legal haggling over Maud's share of the estate was settled in 1919, she had been confined at the Norristown asylum for thirteen years. Maud's residence at the mental institution took place during what has been called the "darkest era" in the history of treatment of the mentally ill. This period was so known not just for the dreadful living conditions in mental institutions; nor for the invasive, quack medical treatments that were forced on patients; but also for an overall societal view that the mentally ill were degenerate human beings.

Maud's survival for so many years owes a small debt to the tradition of humane treatment of the mentally ill that originated in Philadelphia in the nineteenth century. In 1817 a group of Quakers, appalled by the cruel treatment to which those judged insane were subjected, founded the first private mental hospital in America, the Friends Asylum for the Relief of Persons Deprived of the Use of Their Reason. Their method of care became known as "moral treatment" and had an influence on psychiatric medicine that lasted until the darkest era.

The Quakers believed in giving patients private rooms with windows, and in using compassion and conversation to engage them. Their hospitals were set out in the country, where patients were encouraged to stroll throughout the grounds and work the gardens. Patients were involved in decisions about their care, and occupational therapy, horticulture, and arts-and-crafts activities were encouraged. When the Commonwealth of Pennsylvania began building its state hospital system in the middle of the nineteenth century, they adopted many of these moral treatment principles. The Norristown State Hospital was founded in this tradition.

The state hospitals were however all but doomed in their attempts to stay true to moral treatment. They were too large, too understaffed, too poorly funded, and administered by superintendents charged more with managing their budgets than with producing successful treatment outcomes. Additionally, psychiatric and neuropathic doctors argued that moral treatment had no scientific backing, in that it did not identify the causes of mental diseases. These forces rallied against moral treatment and set the stage for the horrors of the darkest era.

Many elements of moral treatment were continued at the Norristown State Hospital: occupational therapy; a large campus with gardens, barns, and fields; a library; arts and crafts; plays and movies; and access to doctors and dentists for physical ailments. Still, the chronic overcrowding and understaffing in the state hospitals after the turn of the century diminished the benefits of these humane features. Administrators started to place a high priority on control and restraint of their wards. In the first decades of the twentieth century, Norristown relied heavily on hydrotherapy. At Norristown, patients were restrained in continuous cold baths for long stretches, or were given needle showers—high-pressure showers of cold water.

Norristown also made extensive use of electrotherapeutics (application of electrical currents to different parts of the body, not just the head) and massage. Their electrical apparatus was crude and not nearly as powerful as that used in electroshock therapy much later in the twentieth century. Norristown continued to use this treatment until the 1920s, when its benefits began to be questioned. Electrotherapeutics survives today as an alternative health option, sometimes combined with acupuncture.

If the Norristown facility escaped the ravages of maltreatment, its overcrowded conditions still offered a fertile environment for epidemic diseases. During Maud's first decade there, the hospital dealt with outbreaks of scarlet fever, typhoid, and tuberculosis. In 1919, the influenza epidemic that caused thousands of fatalities across the nation also hit the hospital, resulting in the death of 9 percent of the residents.

Ironically, it was another scandal that might have saved Maud from some of the worst, most invasive medical procedures used on mental patients in the twentieth century. In 1935 a series of alarming reports on conditions at the Norristown State Hospital caused the governor of Pennsylvania to fire the entire board of trustees. There had been stories of atrocious food for the inmates, of wild parties and lewd behavior by the staff, and of widespread gambling by both patients and staff.

The new board of trustees brought in a highly respected doctor, Arthur P. Noyes, as the hospital's superintendent. Noyes wrote the classic textbook used in training nurses and physicians, *Clinical Psychiatry*, and brought new insights to the study of schizophrenia. He combined scientific and humane approaches in his work. For example, Noyes was skeptical that mental illnesses could be neatly categorized by diagnostic classification schemes, and urged that the patient's entire life story needed to be taken into consideration. During his tenure, Noyes appears to have steered away from involvement in the most heinous so-called treatment that became a fad in the middle of the century: frontal lobotomies.

Under Noyes's care, Maud lived out the rest of her life in the confines of the state hospital. After having been institutionalized for decades, she was an unlikely candidate for any new treatment methods, due to her age and her obeisance to the hospital regimen. Maud's father and stepmother died in the 1930s, leaving her without anyone in the outside world aware of her existence. On May 20, 1947, Maud Maria Lee passed away from myocardial degeneration, heart failure, at the age of seventy-five. She had lived the last forty-one years of her life in the asylum. She was buried in an unmarked plot at the Norris City Cemetery, which bordered the hospital grounds.

A year before Maud's death, a new musical based on the life of Annie Oakley opened on Broadway: *Annie Get Your Gun.* It was produced by Rodgers and Hammerstein and featured songs written by Irving Berlin. The plot of the musical, which centered on the romance between Annie Oakley and Frank Butler, made no pretense at historical accuracy. Had they still been alive, Oakley and Butler might well have considered bringing libel suits against every theater that staged the musical. Liberties were taken with their courtship and with their characters that both would have considered insulting.

In the musical, Annie is portrayed as an uneducated backwoods hick, little more than a hillbilly. Frank is portrayed as a somewhat vain sharpshooter, already starring in Buffalo Bill's show when he meets Annie. He becomes jealous when Annie joins the show and eclipses him as a star, then leaves in a huff to join Pawnee Bill's show. Sitting Bull funds Buffalo Bill's tour of Europe, but the tour is a financial disaster. Buffalo Bill and his troupe arrive back in New York broke and encounter Pawnee Bill's show, also a bust. The shows agree to merge only after Annie allows Frank to win a shooting competition to restore his pride.

The storyline of *Annie Get Your Gun* continues to make historians cringe, because it gets nearly every fact about the Wild West show era wrong. Although Annie and Frank would have hated their public depictions, in their hearts they would not have found complaint with the musical's underlying theme: the triumph of their love for one another. Thanks to Irving Berlin, Annie Oakley and Frank Butler have been enshrined in popular culture as romantic icons. In real life Oakley and Butler guarded their privacy, but there is ample circumstantial evidence that their love for each other was as deep as any Broadway musical fantasy. In that sense, *Annie Get your Gun* does them perfect justice.

As much as *Annie Get Your Gun* celebrates romance, it is also an affectionate memorial to the Wild West show and theater life in general. The musical's finale weaves the appeal of Wild West performing with the love story of the two lead characters

The cowboys, the wrestlers, the tumblers, the clowns,
The roustabouts that move the show at dawn
The music, the spotlights, the people, the towns,
Your baggage with the labels pasted on.
The sawdust and the horses and the smell,
The towel you've taken from the last hotel.

There's no bus'ness like show bus'ness
If you tell me it's so.
Trav'ling through the country will be thrilling,
Standing out in front on opening nights.
Smiling as you watch the theatre filling,
And there's your billing out there in lights.
There's no people like show people;
They smile when they are low.
Even with a turkey that you know will fold,
You may be stranded out in the cold,
Still you wouldn't change it for a sack of gold—
Let's go on with the show.

They say that falling in love is wonderful
It's wonderful, so they say.
And with a moon up above it's wonderful
It's wonderful, so they tell me.[10]

The original cast of *Annie Get Your Gun* was still performing nightly on Broadway in May 1947. As Maud Lee lay dying of a failing heart in a Pennsylvania asylum, Ethel Merman was belting out the Irving Berlin tunes that transformed Annie Oakley into an immortal romantic heroine. Maud had known love once, and it defined her life and aspirations until bad luck, bad judgment, pain, and disease took it all away. She lost her love too soon, but it still served to support her through a career that brought her far more misery than fulfillment. Perhaps, over her last forty-one years, the recesses of her troubled mind still retained the ambitions and emotions that had sustained her in early adulthood, but her illness would not permit her to connect those feelings to the real world. Maud's story is the heartbreaking counterpoint to the sentiment of *Annie Get Your Gun.*

One other event of note occurred in May 1947. Over the decades since his death, the fame of Samuel F. Cody had all but evaporated. Some British aviation historians discounted Cody's contributions to the extent that even his claim of being the first to fly in that country was denied. Two world wars had passed since Cody's pluck and heroics captivated the public, and the massive toll of those conflicts made the daredevil actions of the first aviators in their flimsy machines seem quaint and remote. In post–World War II Britain, few recognized his name. And so one day in May 1947—perhaps the same day that Maud Lee closed her eyes for the last time—the curators of Madame Tussauds wax museum in London removed the statue of Samuel F. Cody from public display and reduced it to its elements.[9]

Through the prism of reflection we can picture the young cowboy Sam Cody, exiting as a free man from the Reading, Pennsylvania, courthouse with a beaming Maud Lee by his side. A moment later we can envision the sad, lonely burial of an old madwoman and the melting effigy of a long-forgotten adventurer. Their ill-fated relationship—containing adventure and spectacle, perseverance and mishap, courage and frailty—was the reality behind the show trappings of the Wild West. Their story contains romance, to be sure, but is also a tragedy of lost potential.

AFTERWORD

Samuel F. Cody and John William Dunne, the military engineer whom many believed would produce the first airplane in Great Britain, never expressed that they were in competition with each other. Dunne might be said to have fared even worse than Cody. Published accounts that appeared immediately after he was let go from the Balloon Factory claimed that Dunne had promised his superiors that he could better the efforts of the Wrights, and for this reason the War Office did not license the Wright plane.[1]

This was plainly untrue—if anyone was to blame for not buying the Wrights' design, it was the penurious Colonel Capper. The *Times* of London also reported that it was Dunne who abruptly severed his relationship with the Balloon Factory, rather than the other way around. There seems to have been a concerted effort to blame Dunne for Britain's lagging efforts in heavier-than-air flight. Dunne vindicated himself in May 1910 when, working independently, he flew his V-wing, tailless machine a distance of 2 ½ miles. In 1914 a Dunne-designed airplane was purchased by the Canadian Aviation Corps as their first aircraft. Lashed to the deck of a ship bound for Europe, it was damaged beyond repair by the rough ocean transit. This episode effectively ended John W. Dunne's aviation career.[2]

Plagued by fragile health, Dunne took up vocations suggesting convalescence: he wrote a children's book, and became an expert fisherman and tied-fly designer. An odd thing occurred to him while in a hospital bed in 1917, however. He read a book that featured an unusual combination lock as a minor plot device—one embossed with letters rather than numbers. The reference bothered Dunne, until he realized that he had dreamed about such a lock the night before. He discussed this coincidence with other patients and discovered that they, too, had pre-cognitive dreams with a frequency that seemed beyond coincidence.

Dunne began keeping logs of his remembered dreams and encouraged others to do so, thereby discovering a pattern of seeing glimpses from the future. Dunne, an engineer by training, sought a theoretical model to explain this phenomenon. His investigation led him to devise a theory of time that suggests that past, present, and future exist simultaneously, but that the conscious mind can only experience one frame at a time. The unconscious mind, however, can freely observe past, present, and future. Dunne therefore concluded that one part of ourselves exists within time, and other parts of ourselves exist outside time—and those parts are, in essence, immortal.

Dunne published his theories in several books, beginning with *An Experiment with Time* in 1926, followed by *The Serial Universe, The New Immortality,* and *Nothing Dies.* Dunne's old mentor, H. G. Wells, wrote glowing reviews of these works.[3] Dunne's mathematical model approach separated his ideas from more metaphysical theories about time, although some followers of mysticism embraced his findings. His theories had a heavy influence on several writers, including J. B. Priestly, Aldous Huxley, and T. S. Eliot. Eliot's poem *Burnt Norton* opens with lines that reflect Dunne's notion of time:

> Time present and time past
> Are both perhaps present in time future,
> And time future contained in time past.

These lines speak directly to the experience of historical writers and researchers who, while immersed in their projects, sometimes develop a special awareness of the past that is as close as may be possible to experiencing timelessness. This feeling is heightened when the writer uncovers details about a person that perhaps the person never told to anyone else or that no one else ever discovered. In this sense, Maud Lee and S. F. Cody helped me travel through time, even though there are undoubtedly many more aspects of their lives and times that would still surprise me.

NOTES

CHAPTER 1

1. Glass target balls are now expensive collectors' items. The search for the origins of glass target balls has become a research quest in itself. See Finch, "Who's On First?" The first documented mention of glass target balls comes from the May 25, 1867, issue of *Bell's Life,* which credits James Harding of Wednesbury, England, with their introduction. The first American mention found so far comes from an article in the *Boston Daily Advertiser* of June 12, 1872, concerning a shooting contest sponsored by the Portland Light Infantry that took place in Portland, Maine. American shooting champion Ira Paine used glass balls and a trap made by Eaton, Holberton & Co. during his exhibition tour in the summer of 1876. In 1877, Adam Bogardus patented his ball design, which added ridged furrows and a more powerful trap to throw them.

2. For an opening-night review of *Si Slocum* that mentions the mirror shot, see the *St. Louis Daily Globe-Democrat,* November 25, 1875, 4.

3. *Brooklyn Eagle,* January 21, 1877. Some accounts claim that Frank Butler performed as one of the two Austin Brothers. However, further research has located them on tours of Brazil and England when Butler was in the United States, so unless Butler filled in for one of them on occasion, this claim can be discounted.

4. Details of Jennie Franklin's career and the fatal incident involving Volante can only be found in contemporary newspaper articles. The shooting accident is described in the *Brooklyn Eagle,* April 6 and 7, 1878, and the *Geneva (N.Y.) Courier,* April 17, 1878.

5. Jennie Franklin's *New York Sun* interview was reprinted in the *Rocky Mountain News,* May 11, 1878.

6. *New York Clipper,* April 27, 1878, p. 1. In addition to giving Butler's view of the accident, this letter reveals new details about Frank Butler's own career, specifically that he spent the summers of 1876 and 1877 managing a shooting range in Gloucester, New Jersey.

7. Both of the most recent scholarly biographies of Annie Oakley—Shirl Kasper, *Annie Oakley,* and Glenda Riley, *The Life and Legacy of Annie* Oakley— mention the discrepancies between Annie's account that her marriage took place

in 1876 and a marriage certificate from Windsor, Ontario, dated June 22, 1882. Kasper also mentions that Frank, on three separate occasions, said that he met Annie in April 1881. Riley is more inclined to accept Annie's date, and posits that Major John Burke, Buffalo Bill's agent, suggested the later marriage date in order to jive with making Annie appear six years younger than she actually was.

Neither Kasper nor Riley had access to digitized, indexed editions of the national theatrical newspaper, the *New York Clipper.* The *Clipper* printed theater act billings from around the country that tell a much different story about the early years of Annie and Frank. In the summers of 1876 and 1877, Butler managed a shooting range in Gloucester, New Jersey. His first recorded stage shooting act (with his trained dog, George) was in November 1877, in Philadelphia. He also appeared at a Camden, New Jersey, theater in December 1878 and January 1879. His first performance with partner Samuel Hyde Baughman took place on July 4, 1879, at the Stockton rifle range in Camden, where Butler worked. In September 1879, the pair played in Baughman's hometown of Baltimore.

The first evidence of Butler ranging beyond the coastal cities is a billing of Baughman and Butler in Bradford, Pennsylvania, in February 1880. In August 1880, nearly two years prior to when Kasper and Riley indicate that Annie first performed, Annie Oakley appeared as a solo act at Brand's Music Hall in Cincinnati. In September 1880, Baughman and Butler were in Detroit; in December of that year, Butler appeared at a shooting contest in Princeton, Kentucky.

In late February 1881, Annie Oakley again played at Brand's Music Hall in Cincinnati; during the same week, Baughman and Butler were billed at the same theater. This is the first documented placement of Oakley and Butler together, and it closely matches Frank Butler's later claim that they met in April of 1881. While this evidence is not conclusive, it does suggest that Annie Oakley started her performing career before she met Frank Butler, and that their marriage really did take place in 1882. Frank Butler's divorce records from his first marriage have not yet surfaced.

8. *Syracuse Morning Standard,* June 9, 1880.

9. *New York Times,* December 12, 1882, p. 1. The shooting took place at the Coliseum Theatre in Cincinnati—the same theater where Frank Butler claimed to have met Annie Oakley.

CHAPTER 2

1. An accurate biography of William F. Carver remains to be written. The one biography of Carver to appear thus far, Raymond W. Thorp's *Spirit Gun of the West*, relies heavily on Carver's own fabrications about his background. In "These Carver Yarns," E. L. Stevenson pokes holes in Carver's reputation as a shooter, and an untitled review of Thorp's book in *Nebraska History* 38, no. 4 (December 1957): 325–26 casts a skeptical shadow on his exploits as a frontiersman.

2. Carver and Bogardus are both enshrined in the Trapshooting Hall of Fame and Museum in Vandalia, Ohio. The Hall of Fame's website hosts a detailed

article by Richard Hamilton on the Carver versus Bogardus shooting matches, "The Sporting Event That Captivated America in the Spring of 1883."

3. Louis S. Warren's excellent interpretative biography, *Buffalo Bill's America,* provides the best account of the beginnings of the Wild West arena show.

4. *Deadwood Dick* was produced by John Burke, one of the managers of Buffalo Bill's Wild West. Despite the swift demise of this production—or perhaps because of it—Burke likely played a role in Oakley and Butler's reunion with Buffalo Bill and in Lillian Smith's departure.

5. This claim is found in Thorp, *Spirit Gun of the West,* and therefore needs to taken with a grain of salt. That Carver would take pride in being named an "Evil Spirit" by American Indians, and in doing so gain the admiration of thousands of whites, is a measure of the relations between the two cultures.

6. The *Oakland Evening Tribune* of December 13, 1876, has a fateful report of a shooting exhibition the previous day by Bogardus at the Oakland Trotting Park. In addition to Bogardus's display of pigeon shooting, the event featured a shooting match among five amateurs, one of whom was Doc Carver. Carver did not do very well that day, and one can imagine the jealousy he felt at the deference paid to the great Bogardus.

7. Lillian Smith began performing in public in San Francisco at age ten, in 1881. Her exhibition included the mirror shot. See the *Butte Daily Miner,* July 30, 1881. By 1884, her father was issuing challenges to Carver and other takers, according to the *Reno Weekly Gazette* of July 31. Even in 1886, when she was with Buffalo Bill's Wild West, Smith continued to issue challenges to Carver, as evidenced in the *Reno Evening Gazette,* May 28, 1886.

8. Forepaugh was only too happy to capitalize on Carver's fame, so the advance publicity that Forepaugh's agents supplied to newspapers on the circus route often contained long, adulatory profiles of Carver. For an example, see the *Newport (R.I.) Mercury,* June 23, 1888, p. 4.

9. The report of the "Jersey Amazon" originated in Philadelphia and was carried by many newspapers. One reprint can be found in the *Dunkirk (N.Y.) Evening Observer,* May 10, 1884, p. 6.

10. The links between Jennie Franklin's various stage names and marriages can be traced from scattered news accounts, but it is likely that only a few within the shooting community made the connection with Jennie's previous identity and killing of Volante. The *Dunkirk Evening Observer,* May 10, 1884, makes an explicit link between Jennie Franklin and Jennie Moore. The *New York Clipper,* September 5, 1885, p. 386, gives a billing of Jennie Moore with W. B. Kennedy and Little Josie. A letter to the editor titled "Petticoats at the Trap," in the March 31, 1887, volume of *Field and Stream,* p. 211, makes the connection between Mrs. W. B. Kennedy and "Mexis." "The Referee" gossip column in the April 2, 1887, issue of the *National Police Gazette,* p. 11, also links the identities of Moore, Franklin, and Kennedy, but does not mention Mexis.

11. The *Bangor (Maine) Whig and Daily Courier* of July 14, 1888, contains an article listing the names of the notable "scouts and girl riders of the far west"

in Forepaugh's Wild West show. Mexis is mentioned first among the women in the cast.

CHAPTER 3

1. Like Doc Carver, Adam Forepaugh is a fascinating and influential character in American popular culture who has been neglected by biographers. Many of the details about Forepaugh in this chapter were taken from an extensive (and critical) profile in the *New York Times,* June 19, 1887, p. 10.

2. The circus poster collection of the Ringling Museum of Art in Sarasota, Florida, includes a depiction of the interior of Forepaugh's winter headquarters in Philadelphia, with a central panel showing the circus riders' training ring.

3. To judge by the advertising space devoted to Forepaugh's 1888 season in various cities, Eclipse and Blondin were the star attractions of the show, more so than Doc Carver or Adam Forepaugh, Jr.

4. No photographs or lithographs of Maud Lee have yet surfaced. This description of her physical appearance comes from a few descriptive phrases found in newspaper accounts of Cody's Reading trial in 1888, and from depositions when she was arrested in 1903.

5. Garry Jenkins's biography of Cody, *Colonel Cody and the Flying Cathedral,* contains some plausible conjectures about Cody's employment with a Texas cattle company, based on a few names and places mentioned by Cody in later years. Much of what Cody said about his early years has however proven to have been fabricated.

6. The *Reading Eagle* account from December 13, 1888, states clearly that Maud had worked as Cody's assistant, holding targets for him.

7. Circus riding was one of the most hazardous occupations in America. These descriptions were taken from articles in the *Waukesha Freeman,* December 13, 1900, p. 3; and the *Philadelphia Inquirer,* August 9, 1895.

8. Maud's stepmother told newspapers in 1913 that Maud and Cody met when he was with Forepaugh's circus in 1888 and she was just a visitor to the show; this contradicts the *Reading Eagle,* December 13, 1888, which states that Maud had assisted in Cody's act in early 1888. The stepmother's account mixes up the sequence of other events, so the Reading version seems more likely.

9. *Hartford Courant,* June 14, 1888, p. 1.

10. *Fitchburg Sentinel,* June 22, 1888.

11. This suspect anecdote of bad blood between Sioux and Apache is found in the *Dunkirk (N.Y.) Observer* of September 12, 1888. Forepaugh's circus came to Dunkirk just six days after the article appeared, so this story was likely planted by his agents to emphasize the bona fides of Forepaugh's cast of Indians and to appeal to longstanding stereotypes.

12. These mishaps were reported in the *New York Times,* July 7, 1888, p. 1; the *Newark (Ohio) Daily Advocate,* September 26 and October 1, 1888; and the *Washington Post,* September 27, 1888.

13. Samuel F. Cody is mentioned by name as a member of the Forepaugh Circus cast in the April 27 issue of the *New York Clipper,* p. 113.

14. *Reading Eagle,* October 12, 1888.

15. Cody's initial hearing and subsequent reaction are recounted in the *Reading Eagle,* October 16, 1888.

16. Cody's trial and Maud's fawning attention are reported in the *Reading Eagle* of December 12 and 13, 1888; and the *Reading Times,* December 13, 1888. The Berks County court docket of December 8, 1888, lists the charge more explicitly than the newspaper accounts: assault with intent to feloniously rape.

17. The *Philadelphia North American*'s December 25, 1888, review was only slightly more positive: "bound to please even the exacting tastes of the youths who every now and then start west with the avowed purpose of increasing the supply of dead Indians. . . . Miss Annie Oakley was acceptable as Jeannette."

18. Documentary evidence proving that S. F. Cody was in the cast of *Deadwood Dick* has not surfaced, but Cody's claim that he was in the show fits a blank period between other documented events, and he was in the vicinity (Reading beforehand, New York City afterwards). Moreover, many years later Maud Lee gave a deposition in which she said she had met Annie Oakley in the early months of 1889, which also fits.

19. *New York Clipper,* February 9, 1889, p. 767.

20. The acts are listed in the *New York Clipper,* February 16, 1889, p. 781.

21. *New York Clipper,* March 16, 1889, p. 5.

22. The only other account of Harry Hill's Oklahoma Historical Wild West Exhibition that has been published is in Hoig, *Cowtown Wichita and the Wild, Wicked West.* Mr. Hoig graciously shared his chapter on Harry Hill with me shortly before his book appeared in print.

CHAPTER 4

1. The Cherokee nation's objection to the opening of Oklahoma was expressed in a petition to Congress, as reported in the *Galveston Daily News,* February 26, 1889.

2. See Hoig, *Oklahoma Land Rush of 1889,* for a summary of the political maneuvering behind the boomer movement, including the involvement of Pawnee Bill.

3. The communication from Washington to Hill in Kansas, and Pawnee Bill's plans for imminent invasion, are reported in the *Galveston Daily News,* February 3, 1889, and the *San Antonio Daily Light,* February 2, 1889, p. 1.

4. *Galveston Daily News,* February 8, 1889, p. 2.

5. *Rocky Mountain News,* February 26, 1889.

6. *Galveston Daily News,* March 11, 1889, p. 1.

7. *Van Vert (Ohio) Republican,* March 28, 1889.

8. The marriage was recorded in the *Philadelphia North American,* April 10, 1889; and the *Philadelphia Inquirer,* April 11, 1889.

9. *Chicago Tribune,* April 12, 1889.

10. "Bronco Charley" may have been "Broncho" Charlie Miller, a veteran cast member of Buffalo Bill's Wild West show. He is mentioned in a review of the Oklahoma Historical Exhibition's Chicago show in the *Daily Inter Ocean,* May 28, 1889, p. 7. "Bronco Charley" is not however mentioned in other articles about Harry Hill's show, and Charlie Miller did not mention the show in his autobiography, *Broncho Charlie,* transcribed by Gladys Erskine.

11. Maud's assistance in young Harry Hill, Jr.'s, shooting act is mentioned in the *Daily Inter Ocean,* May 28, 1889, p. 7.

12. In the early 1880s, when Wichita was a boomtown, Harry Hill, David Payne, General William Babcock Hazen, Dave Bancroft, and William Couch played a legendary game of poker. They started by playing for cash, but when Hazen had accumulated all the available money, the players began to wager with Wichita town lot deeds. This story was written down years later and reprinted in the *Lima (Ohio) Times-Democrat,* April 19, 1900.

13. Side-by-side ads for the Oklahoma Historical Wild West Exhibition and Paul Boyton's exhibition can be found in the *Kansas City Star,* May 10 and 11, 1889. Boyton gained fame by crossing the English Channel and floating down the major rivers of the world in a Merriman suit—an early wetsuit. Many years later, Cody would make his own spectacular channel crossing.

14. Harry Hill, whose death had been mistakenly reported on more than one occasion, did pass away on July 10, 1896, after falling from a buggy and being hit in the head by a mule. The death notice appeared in the July 22, 1896, edition of the *Fort Wayne (Ind.) Morning Journal* and the *Macon (Ga.) Telegraph,* July 17, 1896, p. 3.

CHAPTER 5

1. Maud Lee mentions being in this show on page 3 of a deposition she gave many years later. Using her stage name Lillian Cody, she was deposed in Norristown, Pennsylvania, on October 30, 1905, as agreed to by the parties in the case of *Butler v. Carter & Russell Publishing Co.,* U.S. Circuit Court for the Southern District of Florida, 1905. The court decision appears in the *Federal Reporter,* and the depositions can be found in the collection of the Annie Oakley Center at the Garst Museum, Greenville, Ohio.

2. *New York Clipper,* January 18, 1890, p. 753.

3. Maud's accidental shooting of Mary King was reported to news services and printed in newspapers all around the United States. For examples, see the *Decatur (Ill.) Morning Review,* May 14, 1890, p. 1; and the *Bucks County (Pa.) Gazette,* May 15, 1890.

4. Cody's date of departure from the United States was noted in his 1892 passport application. His booking at St. Helens is found in the *London Era,* July 12, 1890, p. 17.

5. *London Era,* July 19, 1890, p. 17. The ads the Codys placed to secure additional engagements appeared in the same paper on July 26 and August 2, 1890.

6. The Codys' appearance in Birmingham was noted in the *London Era,* September 6, 1890

7. These names were found by scanning digitized copies of the *London Era* for that period. The *Era* contained notes of current bookings at theaters throughout Great Britain, and also printed advertisements placed by performers and their agents looking for additional bookings.

8. Detailed descriptions and photographs from each of Whitley's four shows can be found in Lowe, *Four National Exhibitions in London.* The London newspapers also printed very complete program details in the paid advertisements for the exhibitions. The French Exhibition events are found in the *London Daily News,* November 1, 1890, p. 4. The attractions at the French Exhibition were reviewed in the *Times (London),* May 19, 1890, p. 7.

9. The first reviews of the Wild East arena show can be found in the *London Daily News,* May 22, 1890; and the *Pall Mall Gazette,* May 22, 1890, p. 6.

10. A complete daily program of the Wild East show is reprinted in Lowe *Four National Exhibitions in London,* fn. on pp. 264–65.

11. See the review in the *Pall Mall Gazette,* May 22, 1890, p. 6.

12. The reference to the King of Dahomey and his Amazon bodyguards is found in the *London Daily News,* May 6, 1890, p. 6. Mention of the new Jules Verne underwater show found its way into the *Manitoba Daily Free Press,* October 25, 1890.

13. Following their engagement with the French exhibition, the Codys advertised for billings at other theaters. In the *London Era,* November 8, 1890, p. 26, they reprinted a letter of reference from Victor Applin, the secretary of the exhibition, praising their work. This letter names the roles that the Codys played in the Wild East show.

14. A delightful sketch of the skating rink at Olympia from the *Pall Mall Budget,* April 24, 1890, is reprinted in this book.

15. Hall put his name on the Wild West burlesque, but the script was written by Edward Saucier, according to the review in the *London Daily News,* December 27, 1890, p. 6.

16. Terrible puns were in vogue in London of the 1890s. One example was offered on the occasion of the Putney show. From the *London Era,* May 23, 1891: "'What is the difference between Professor Charles Baldwin and Captain Cody and his sister?' asked a wag. 'Give it up, eh? Well, one is a parachutist and the others are a pair-of-shootists.'" The *London Era,* May 23, 1891, reported on Charles Baldwin's parachute jump. No accidents nor other jumpers were mentioned, although Maud's family later claimed she was injured during a parachuting accident.

17. The *Times (London),* July 4, 1891, p. 5, reports on the case against Frank Albert being brought before the High Court of Justice, Chancery Division.

18. Whitley's speech and Blücher's response are reprinted in Lowe, *Four National Exhibitions in London,* 319–23.

19. *Amusing Journal,* July 8, 1893.

20. Maud's involvement in teaching Vivian how to shoot was stated by Maud's stepmother/aunt after Cody's death. See the *Camden (N.J.) Post Telegram,* August 12, 1913. The fact that Maud was understood to have claimed Vivian as her own child in the muddled account heard by her jailers in 1903 affirms that she fixated on Vivian as an emblem of her betrayal.

21. *Birmingham Daily Post,* October 24, 1891. The closing performance of "Hall's Burlesque of the Wild West" is announced in the November 7 edition of the same paper.

22. The November 8, 1891, passenger list of the *Scythia* is available at ancestry.com.

23. After Cody's death, Maud's stepmother/aunt described Maud's pitiful condition upon her arrival home in the *Camden Post Telegram,* August 12, 1913.

Chapter 6

1. See Maud Lee's October 30, 1905, deposition in *Butler v. Carter & Russell Publishing Co.,* p. 3.

2. See the *Camden Post Telegram,* August 12, 1913; and the *Philadelphia Inquirer,* August 12 and 13, 1913, for the Lees' statements critical of Cody's treatment of Maud.

3. Joseph, Phoebe, and daughter Maud can be found living at Ashton-under-Lyne in the 1881 British census. In the same census, Elizabeth Nadin, age twenty-three, is listed as living with her mother, but an Elizabeth Kay, age twenty-four, is married to William Kay. In the 1910 U.S. census we find Joseph Lee living in Swedeland with his sister-in-law, Elizabeth Kay, confirming that Elizabeth Kay was the former Elizabeth Nadin. This suggests that Elizabeth was counted twice in the 1881 British census, before and after her marriage.

4. Maud's 1872 birth record from Ontario is available at www.ancestry.com (accessed March 11, 2008).

5. In the 1920 U.S. census, Joseph Lee reports he immigrated to the United States in 1884. Phoebe and Maud Lee arrived at Philadelphia from Liverpool on August 3, 1885, on the *British Princess,* according to the passenger list. Joseph was not traveling with them, which supports the idea that he had arrived the previous year.

6. Joseph and Elizabeth Lee are found living in Camden in the 1920 and 1930 U.S. censuses.

7. Maud's death certificate indicates that she had an underlying condition of dementia praecox. This is the only documentation that identifies Maud as a likely schizophrenic. All medical records of the Norristown State Hospital are sealed by confidentiality laws of the Commonwealth of Pennsylvania and can be opened only by a court order instituted by a direct descendent. Maud had no children.

8. The first mention of Maud's addiction is found on pp. 23–24 of the October 25, 1905, deposition of Anna Murphy in *Butler v. Carter & Russell Publishing Co.* It is not known when Maud began using morphine.

9. Ads for the Cody family act can be found in the *London Era*, April 9, 1892; and the *(London) Graphic*, March 26 and April 16, 1892. The Codys' frequent bookings at the Royal Aquarium in 1892–93 suggest that they must have been on very good terms with the managers.

10. Maud related the confrontation with Lela King to police matron Anna Murphy when she was jailed in August 1903. The account appears in the October 25, 1905, deposition of Anna Murphy in *Butler v. Carter & Russell Publishing Co.*, pp. 23–24.

11. In the *Leeds Mercury*, November 28, 1891, there is an ad for the "Cody Family—Marvels—The World's Greatest Fancy Shots," appearing at the Fenton Street Drill Hall skating rink.

12. Earlier biographies of Cody by Broomfield (*Pioneer of the Air*) and Lee (*Flying Cathedral*) repeat Cody's fabricated story. The death of John Blackburn Davis, seventy-three, residing in Chelsea, is noted on Ancestry.com, England & Wales, FreeBMD Death Index: 1837–1983, 1890, Jul–Aug–Sep, p. 76.

13. Riley, *Life and Legacy of Annie Oakley*, 74. It is tempting to think Butler might have seen the Codys' act while in England in 1892, but this quotation appears to be from the first trip made with Buffalo Bill's show in 1887.

14. The Codys' 1893 billings at the Royal Aquarium are documented in *Pall Mall Gazette*, July 3, 1893, p. 1; *Graphic*, July 8, 1893, p. 35, and August 19, 1893, p. 215; *Reynolds's Newspaper*, July 23, 1893, p. 8, and July 30, 1893, p. 4; and *Lloyd's Weekly London Newspaper*, August 6, 1893, p. 7. The voyage to France is documented in Reese, *Flying Cowboy*, 26.

15. Cody's stepson Vivian is the source of this anecdote.

16. *Times (London)*, October 30, 1893, p. 10.

17. *Times (London)*, November 14, 1893, p. 12; and *New York Times*, December 20, 1893, p. 6.

18. The terms of the race in Bordeaux were published in the *Oshkosh (Wisc.) Daily Northwesterner*, December 22, 1893.

19. *Times (London)*, August 20, 1894, p. 4.

20. The first mention of Cody in the pages of *Cycling* appears on January 6, 1894, p. 457. By August 6, 1898, the Catford Cycling Club felt compelled to print a disclaimer stating that the course where it raced had been leased to Cody by the landowners, and the club had nothing whatsoever to do with Cody.

21. Page 3 of Maud's deposition in *Butler v. Carter & Russell Publishing Co.* contains two mysteries. She states being with the "B. & B. Wild West," which does not match the title of any known Wild West show—unless she was referring to the Buckskin Bill Wild West show, which she mentions in the next breath. She also mentions performing in Germany, France, Italy, and Mexico; no record of these venues has been found.

22. See the sports pages of the *Omaha World Herald*, August 15 and 16, 1894.

23. An ad in the *Waterloo (Iowa) Daily Courier,* August 24, 1894, for Lillian Cody's appearance at the Iowa State Fair claimed that she had been "Buffalo Bill's best rifle and revolver shot" and had been with the "World's Fair Wild West Show," a reference to Buffalo Bill's Wild West show at the 1893 Columbian Exposition in Chicago. This is an example of the false publicity claims made for Maud; she later claimed agents, not she, were responsible for them.

24. *Annual Report of the Iowa State Agricultural Society* (1894), 20.

25. *Thirty-Fourth Annual Report of the Secretary of the State Board of Agriculture.*

26. Maud's bold offer to ride any unruly horse brought before her was made in advance of both the Iowa and Texas state fairs. The Texas challenge was reprinted throughout the country; see, for example, the *Newark (Ohio) Daily Advocate,* August 23, 1894, p. 8. When the Texas State Fair's attractions were described in the *Galveston Daily News* of September 10, 1894, p. 7, however, only Doc Carver's name is mentioned.

Chapter 7

1. While appearing at the Circus Lenka in Ghent, Cody inspired the formation of the Royal Cody Ghent Rifle Club, which is still in existence. The newsletter of Olympic shooting sports in Belgium, *De Olympische Schietsport,* no. 3 (2004): 11, mentions the origins of the club.

2. The first biographies of Cody imply that Cody invented a machine pistol and do not mention the Borchardt by name (see Lee, *Flying Cathedral,* and Broomfield, *Pioneer of the Air*). The gun that was sold in the auction of S. F. Cody's possessions in 1996 was definitely a Borchardt C-93; and it is this gun that Cody is pictured with in photographs. Cody may have modified the Borchardt, but he did not invent it.

3. There is not much in the way of documentary evidence to assess the character of Lela King, S. F. Cody's common-law wife. However, she was not timid about displaying confidence in Cody: she held his targets in the shooting act for many years, she went aloft in his man-lifting kite, and she agreed to be a passenger in his earliest airplane flights.

4. In the *London Era,* January 17, 1891, Cody and Maud advertised that they were available for bookings, and that their act included a sketch called "Always on Hand." They both had apparently enjoyed their time in the Wild West burlesque, which may have inspired them toward theatrics.

5. The August 2, 1897, Wood Green bicycle meet that Cody promoted drew 16,000 spectators, according to the *London Daily News,* August 3, 1897, p. 6. Cody, Lela, and Leon each raced bicyclists with their horses. Cody also raced astride two horses and in a chariot. There were however also legitimate bicycle-only races in the event.

6. The performances by the Cody family at Alexandra Palace during the summer of 1898 were billed as "Colonel Cody's Wild West Show" and featured

representations of "numerous stirring events." See the *Times (London),* June 3 and July 12, 1898; and the *London Era,* June 4, 1898, p. 14.

7. In early August 1894, during an ascent at Folkestone, Gaudron's balloon was driven by wind out over the sea, and he was compelled to parachute into the water and await rescue by boatmen. See *Jackson's Oxford Journal,* August 11, 1894, p. 2.

8. *Lloyd's Weekly London Newspaper,* August 2, 1896, p. 15; *Bristol Mercury,* July 28, 1896, p. 8.

9. According to Broomfield, *Pioneer of the Air,* Gaudron and Cody shared the same workshop space at Alexandra Palace.

10. "Historical Cincinnati" was the name of a series of skits presented by the Sioux and other performers at the Cincinnati Zoological Gardens for a few weeks during the summer of 1896. The entire summer's entertainment schedule apparently was not given any specific thematic title, however, being called in ads simple the "Sioux encampment."

11. Vic F. Cody's real name, Anderson, was revealed under questioning during Maud's 1905 deposition in Norristown. The story of how Anderson acquired his stage name is also found in her deposition. Victor F. Anderson, circus actor, is found in the 1930 census living in Pawnee County, Nebraska.

12. Maud's uses the word "assistant" in her deposition to describe Vic Cody; she seemed to want to emphasize that he was never her partner.

13. The John C. Weber Military Band was a Cincinnati fixture. Five years later, in 1901, they were a featured act at the Pan-American Exposition in Buffalo, New York. In Buffalo, they performed "The Civil War," a "realistic tone picture of the bloody scenes in the war of rebellion." The combination of Wild East, represented by Sie Hassan Ben Ali's Arabs, and the Wild West, represented by the Sioux, was reported in the *Enquirer (Cincinnati),* June 28, 1896, p. 19.

14. The gift of the St. Bernard pup to Maud and her "pleasure trip" to the Rosebud Reservation were noted in the *New York Clipper,* October 31, 1896, p. 553. The image of Annie Oakley posing with her St. Bernard, Sir Ralph, is owned by the Buffalo Bill Historical Center. It is reprinted in this volume.

15. S. F. Cody used composition, or resin, target balls in his shooting act. Sotheby's 1996 auction of the S. F. Cody Archive included Cody's ball molds. See Sotheby's, *S. F. Cody Archive Sale,* auction catalog.

16. Maud's recipe for target balls is not known, but target ball collector Ralph Finch (e-mail to author) provided the text of this 1883 patent:

George Westerman Jr. of Lockport, New York, and William T. Miller of Buffalo, New York, for a Composition for Target Balls, Patent number 290,296, dated Dec. 18, 1883, filed Sept. 27, 1883. Westerman and Miller patented a hollow ball made of coal tar "commingled with coal ashes or pulverized fire clay." The recipe for this mixture is: 60 pounds of coal ashes which have been finely sifted. Place in a kettle or other suitable vessel. Subject to heat until the tar is in a molten state and when the ashes can be thoroughly commingled with the tar by stirring. Pour into a series of hollow molds. Wait until a crust or shell has been formed to the desired thickness. Invert the molds and pour out the interior liquid, forming a hollow globular shell.

17. Maud's reference to a formula for target balls was mentioned in the *Anderson (Ind.) Daily Bulletin,* January 6, 1897; the newspaper uses the word "receipt," but from the context of Maud's remarks she clearly meant "recipe."

18. *Indiana State Journal,* November 18, 1896, p. 7, reports the arrest of "Lillie Cody." Later newspaper reports on her from Anderson, Indiana, mention that she had just come from Indianapolis.

19. *Anderson (Ind.) Daily Bulletin,* January 6 and 9, 1897.

20. *London Daily News,* May 31, 1898, p. 8; *Times (London),* May 31, 1898; *London Era,* June 4, 1898, p. 14.

21. The Blue Hill Observatory's record kite-flying experiments reached the British popular press at an early date. See *London Daily News,* December 1, 1896, p. 8.

22. Apocryphal stories about what inspired Cody to take up kite flying can be found that attribute his interest to one of his stepsons, either Leon or Vivian. One account says that Cody was impressed that a group of boys flocked around one of his boys after he purchased him a parrot-shaped kite. These stories do not explain how Cody could have graduated from commercial toy kites to large-scale scientific kites, however.

23. *Times (London),* February 19, 1898, p. 10.

24. *London Daily News,* August 29, 1898, p. 7.

25. The complete script of *The Klondyke Nugget* and partial scripts of Cody's other plays were sold in the 1996 Sotheby's auction and now reside in the archives of the Autry National Center in Los Angeles, Calif.

26. A review of the opening of *The Klondyke Nugget* can be found in the *London Era,* December 10, 1898, p. 9. The play, it noted, "bids fair to have a very successful career, abounding as it does in effective scenes—novel, even in these days of realism—startling situations, and hairbreadth escapes."

27. There is no concrete evidence that Cody and Maud ever communicated after Maud left England in November 1891. On the other hand, there is also nothing in Cody's later career that would indicate that he was a hard-hearted man—just the opposite. Little hints exist which could be used to argue that perhaps they did have contact: Maud's claim that "Buffalo Bill" paid to place her in sanitariums; Maud's remark about Cody drowning on the way to the North Pole, apparently derived from a statement Cody made to newspapers indicating his interest in using kites to power sleds for polar exploration; and Cody's use of the name Lee in *The Klondyke Nugget.*

CHAPTER 8

1. The St. Helens fire was reported in the *Liverpool Mercury,* October 14, 1899. Cody's property losses, estimated at £250, were not covered by the theater's

insurance. *The Klondyke Nugget*'s reopening in Liverpool was announced in the *Liverpool Mercury,* October 27, 1899. The speed with which Cody got his company back in action presages the herculean effort he made to build a new airplane and present it for the 1912 Military Trials in a two-week period.

2. Many of the dates of Cody's kiting activities during the 1899–1902 period have been compiled by Cody researcher Jean Roberts within a comprehensive chronology of Cody's life. Roberts has not published this extensive document, the result of years of research; but she did permit a German kiting hobby site to publish a section of it, entitled "Cody—Diary."

3. An image of one of the Barton-Gaudron models appears in the *Watertown (N.Y.) Daily Times,* March 4, 1901, p. 9. Barton describes the models in chapter 6 of his autobiography, *Jack of All Trades.*

4. In the 1900 census, Maud gave her name as Maud M. Cody and reported that she had been married for eight years but was not divorced.

5. The formation of the Cody Sisters made the theater gossip news in the *National Police Gazette,* April 5, 1902, p. 2.

6. For Fletcher Terrell's background see Robertson, *Paducah.* No proof has surfaced, but Maud might also have been with the Buckskin Bill show in 1901. The casting call for the 1901 season appeared in the *New York Clipper,* January 12, 1901, p. 1031. Ironically, the same page of the *Clipper* has a large advertisement for a treatment for morphine addiction being offered by a philanthropic organization called the St. James Society; their treatment was heroin, and they gave away free samples through the mail.

7. W. F. "Harry" Brandon is cited as playing the role Buckskin Bill in a review from the *Detroit Free Press,* June 11, 1900. Victor F. Cody's stint as Buckskin Bill in 1901 is confirmed in several advertisements and press releases; see, for example, *Athens (Ohio) Messenger and Herald,* May 30, 1901, p. 4.

8. The sale is mentioned in the *Hamilton (Ohio) Evening Democrat,* February 15, 1902. The editor could not restrain himself from adding that the 1901 version of the show had been "the worst thing that ever struck Hamilton."

9. *Dallas Morning News,* September 23, 1902.

10. Maud's injury was reported in several newspapers; see, for example, "Woman Hurt," *Paducah Sun,* May 16, 1902; "Wild West Performer Hurt," *Hopkinsville Kentuckian,* May 20, 1902, p. 5.

11. See Hornaday, *Extermination of the American Bison.*

12. The principal human acts were Vic Cody and Lillian Cody (Maud); Al G. Barnes and his dog-and-pony show; Professor Wheeler's Famous Concert Cowboy Band from Portland, Oregon; Miss Austin, a sharpshooter; Mr. Tompkins, a trick rider; and Frank M. Chamberlain, a "rope juggler" (i.e., lasso artist). The *Bismarck Tribune* of Wednesday, July 9, 1902, in its announcement of the Wild West show, gave more mention to buffalo rancher Charles Allard, Jr., than to any of the performers. By August, the publicity materials supplied to newspapers mentioned only Lillian Cody and Professor Wheeler. Hutchin's quotation appears in the *Waterloo (Iowa) Daily Reporter,* August 29, 1902, p. 4.

13. The assembly of performers in advance of the opening of the Great Buffalo and Wild West Shows United is reported in the *Daily Missoulian,* June 18, 1902, p. 3. Vic Cody and Maud are both mentioned in this article.

14. Most of the pictorial ads and publicity articles for the Great Buffalo and Wild West Shows United mention that the opening parade in each town would be led by Lillian Cody and Professor Wheeling; see, for example, the *Dubuque Telegraph-Herald,* August 30, 1902.

15. See the *Daily Missoulian,* July 3, 1902, p. 1.

16. The entire story of the show, including the negative reviews, can be found in Firman H. Brown's article, "The Great Buffalo and Wild West Show, 1902," in Ogden, McDermott, and Sarlós, *Theatre West.*

17. The alarming mortality rate of the show's bison herd is noted in the *Daily Missoulian,* September 12, 1902.

18. *Iowa State Reporter,* September 12, 1902.

19. On the disgust of the Montana roustabouts, see Brown, "Great Buffalo and Wild West Show."

CHAPTER 9

1. Cody's kite experiments drew the attention of not only the aeronautical research community, but also the general scientific community. An article titled "Observation War Kites" about Cody's works appeared in *Scientific American.*

2. Accounts of the kite-flying contest in Worthing were reprinted throughout Britain and America. An example is the *Washington Post,* August 3, 1903, p. 5.

3. Cody, "Kite That Lifts a Man." *Pearson's Magazine* 16, no. 8 (1903): 106–13.

4. Cody's ambition of personally launching an expedition that used kite-driven sleds to reach the North Pole is found in the *Newcastle Chronicle,* July 18, 1902. I have not yet found evidence that this information was ever published in America, yet Maud Lee in her 1905 deposition maintained that S. F. Cody had died while headed to the North Pole. Someone who knew Cody in England, or Cody himself, must have informed Maud.

5. Barton, *Jack of All Trades.*

6. Cody first proposed crossing the Irish Channel in early 1903, but never made that attempt. His focus turned to the English Channel later in the year.

7. See the *Guardian,* November 9, 1903, p. 10.

8. Many of the details about the Forepaugh-Fish Wild West Show were taken from Georgian, "Luella Forepaugh-Fish Wild West Show, 1903." Richard A. Georgian is the grandson of the leader of the Cossack act, Alexis Georgian.

9. Many newspaper accounts asserted that Maud gave her name as Annie Oakley. The claim was repeated in several lawsuits. Every contemporary account may have missed the possible nuances. The confusion over her identity could have arisen in several ways: (1) Perhaps Maud related her history and her role of

"Any O'Klay" in the 1890 London burlesque, and her listeners confused that role with the English appearances of the real Annie Oakley. (2) The arresting officer, Mulcahy, thought her to be Annie Oakley. Did he hear someone else refer to her as *an* Annie Oakley, or did he hear from Maud's lips the claim that she *was* Annie Oakley? (3) Did Maud herself believe that "Annie Oakley" was a fictionalized character, like Buckskin Bill, who was portrayed by several different actors? Did she realize that only Mrs. Frank Butler could claim that identity? After hearing her deposition, Frank Butler concluded that it was the newspaper reporters and editors who were responsible for the misidentification. He viewed Maud as a pitiful imitator, not an impostor.

10. The facts of what transpired during Maud's arrest have been pieced together from the depositions of Lillian Cody, in Norristown, Pennsylvania, on October 30, 1905; and Anna Murphy and John R. Caverly on October 25, 1905, in *Butler v. Carter & Russell Publishing Co.* The depositions are in the collection of the Annie Oakley Center at the Garst Museum, Greenville, Ohio.

11. Pratt's newspaper account of Maud's arrest, reprinted here, is found in the *Ottumwa (Iowa) Courier,* August 11, 1903, p. 8. Many other newspapers abridged the story to fit their page layout. George W. Pratt was also deposed for one of the libel trials. His deposition can be found in the collection of the Annie Oakley Center. Pratt is contradicted on many points by the testimonies of Maud, Anna Murphy, and Judge Caverly.

12. The example headlines come from the *Omaha World Herald,* August 11, 1903; the *Ottumwa Courier,* August 11, 1903, p. 8; and the *Decatur Morning Review,* August 11, 1903.

CHAPTER 10

1. *Nevada* was performed to establish copyright at the Carlton Theatre in Birmingham, England, on January 8, 1904, according to the *New York Clipper,* April 23, 1904, p. 205.

2. In his autobiography, Barton relates

One morning I had been up to the City in top hat and frock coat, as was usual in those days—and was walking up the hill [at Alexandra Palace] to go back to my shed when Cody, who had a flight of kites up, said: "Come on, Doc, go up." I told him I could not do so in these clothes, when he shouted: "Damn the clothes, go up you———!" using a term of endearment common amongst sailors.

Taking off my hat and coat, I got on the narrow seat and in a short time was about 300 feet up in the air, swaying from side to side over a wide area, as there was a strong gusty wind blowing.

I soon had enough of it, as it was very cold up aloft so I pulled on the cord to raise the back of the kite. I pulled and pulled until I almost lifted myself out of the seat but nothing happened, then suddenly the tail went up and I went down the cable like a streak of lightning.

I managed to check the kite just as it reached the ground and landed safely, but it certainly was one of the most thrilling experiences in my life. (Barton, *Jack of All Trades,* 74–75)

3. Barton's autobiography has two chapters that relate to his airships. In chapter 7, he explains that the explosion that injured him occurred as they were generating hydrogen from treating steel filings with sulphuric acid and water. Barton was burned both by flame and by acid. Defying his colleagues, Dr. Barton was back at work within five days.

4. *Observer (London),* July 23, 1905, p. 5. Barton's airship carried four men: Barton, Auguste Gaudron, Harry Spencer, and Frederick L. Rawson. Rawson, an engineer, assisted with the design, and had an interesting later career as a Christian Scientist and leader of the metaphysical New Thought movement.

5. *(Edinburgh) Scotsman,* October 7, 1907, p. 7.

6. The amount that the Butlers sought in each libel case was published in Stotesbury, ""Famous 'Annie Oakley' Libel Suits." Stotesbury represented one of the defendants, the Publishers' Press Association news service, and so might have wanted to publish the damages being sought in an effort to portray the Butlers as mining the case for gold.

7. See *Waterloo Daily Courier,* August 30, 1946; *Washington Times,* January 10, 1906, p. 9; *Hartford Courant,* October 27, 1905, p. 8; and *Chicago Tribune,* August 12, 1903.

8. In her 1905 legal deposition, Maud admits to having known Annie Oakley "a couple of weeks before I was married," i.e., in early 1890, when S. F. Cody was in the play *Deadwood Dick.*

9. The marriage application of "Maude M. Cody" and "Gartano Fontunillo" is held in the Montgomery County Archives, Norristown, Pennsylvania, and dated April 24, 1905. It lists the names of Maud's parents, confirming her identity.

10. The examples of overcrowding were supplied by the hospital administration in the *1906 Annual Report of the Norristown State Hospital,* on file at the State Library of Pennsylvania, Harrisburg.

11. The Wilkes-Barre paper's editorial was quoted in an editorial entitled "The Shame of the State" that ran in the *Philadelphia Inquirer,* April 21, 1906.

12. The visit was reported in the *Philadelphia Inquirer,* December 30, 1906.

13. The humane aspects of care, and the demographics of the residents, were also reported in the *1906 Annual Report of the Norristown State Hospital.*

CHAPTER 11

1. So negative was Britons' self-image of their country's aviation progress that Cody's landmark flight made news only for its crash at the end. "Army Aeroplane Wrecked," was the headline in the *Guardian* on October 17, 1908, p. 5. Only in the third paragraph is it mentioned that the machine flew five hundred yards.

2. The *Guardian,* in keeping with its attitude from the previous fall, ran the headline "The Army Aeroplane: Another Smash," on January 21, 1909, p. 7.

3. The *Times (London),* May 15, 1909, p. 10.

4. The *Times (London),* June 19, 1909, p. 6, quotes Cody as blaming the oscillating flight on lack of power; but although Cody later admitted he wished for a more powerful engine, he explained that the oscillation was caused by his testing of the controls.

5. Penrose, *British Aviation,* 181.

6. The first flights of Colonel Capper and Lela Cody as passengers are reported in the *Guardian,* August 16, 1909, p. 6. Some papers, for instance the *Mansfield (Ohio) News,* September 16, 1909, incorrectly reported that Lela piloted the airplane by herself.

7. The *Guardian* ultimately changed its tune on Cody's accomplishments. On September 9, 1909, it greeted his record forty-mile flight with the headline "Mr. Cody's Success."

8. American newspapers reported on Cody's naturalization as if it were an affront to his home nation. Still, the *Auburn Citizen* article of October 22, 1909, ended with, "Then catching a glimpse of the Stars and Stripes floating over his shed, he said with some emotion, 'While I am proud to be an Englishman, I am no less proud of the country that gave me birth. I won't haul down Old Glory. She can stay up.'"

9. The United Press news service reported that Cody was as good as dead. "Aeroplane Dashed to Earth: Capt. S. F. Cody, American in England, Was Fatally Injured near Aldershot This Morning," was how the story ran in the *Dunkirk (N.Y.) Evening Observer,* on June 10, 1910, p. 1.

10. The *Observer (London),* August 6, 1911, hailed Cody's finishing the race, saying that he had "pluckily persevered."

11. The idea of Cody going to Australia is reported nearly a year later in the *Observer (London),* September 8, 1912, but is twisted to appear that the Australian government was courting Cody, not the other way round. By the fall of 1912, Cody was riding a series of successes that might indeed have caused the Australians to reconsider.

12. News of the Military Aeroplane Trials and its results were closely followed by the public. The *Observer (London),* September 1, 1912, reports that on a visit to Salisbury to give a speech shortly after the trials, "[Cody] was met on the outskirts of the city by the band of the Wiltshire Territorial Infantry, and was greeted by cheering crowds as his motor car proceeded on the way to the municipal buildings, where he was given an official reception by the Mayor and Corporation."

13. The *Observer (London),* March 9, 1913, makes it clear that Cody had in mind a series of increasingly large seaplane designs. Other editorials and speeches that Cody gave during this period also indicate he was aware that airplanes would become instruments of war.

14. Cody's death was reported in the *Times (London)*, August 8, 1913, p. 9, with a grave attention to detail.

15. See the *Observer (London)*, August 10, 1913, and the *Guardian*, August 11, 1913, for the results of the initial inquest into Cody's death.

16. The funeral, "remarkable in many ways," was described in detail in the *Guardian*, August 12, 1913, p. 7.

17. The *Times (London)*, August 9, 1913, p. 8.

CHAPTER 12

1. Joseph and Elizabeth Lee spoke to newspapers a few days after Cody's death, telling their side of Cody and Maud's broken marriage and announcing their intentions to sue for his estate. See the *Philadelphia Inquirer,* August 12 and 13, 1913; and the *Camden (N.J.) Post Telegram,* August 12, 1913.

2. Cody researcher Jean Roberts has a copy of the program of the Cody benefit, listing all the performers and their sketches. It is a wonderful artifact of British theater circa 1913.

3. Dolly Shepherd, a female aeronaut hired by Auguste Gaudron, cowrote a memoir entitled *When the Chute Went Up.* Shepherd, before Gaudron hired her, waitressed at the refreshment stand at Alexandra Palace and once served a table where Gaudron, Cody, and John Philip Sousa sat. Shepherd also once substituted for Lela Cody, holding targets in Cody's sharpshooting act. It appears Cody and Shepherd enjoyed a mild flirtation, but by her account it was nothing more.

4. According to one unconfirmed source found on the Internet, both Gaudron and Percy Spencer are buried in Highgate Cemetery, Highgate, London. Both are said to have gravestones sculpted as hot-air balloons.

5. In early 1913, Cody advocated the use of airplanes as anti-dirigible weapons. He devised a scheme whereby passing airplanes could drag grappling hooks fitted with explosive tips along the sides of airships. See the *Washington Post,* April 27, 1913, p. MS1.

6. When World War I began, Frank Cody was working as a designer at the factory of the Royal Flying Corps. He preferred flying, though, and was sent to the front. His death did not make headlines until weeks later. See the *Great Falls Tribune,* May 4, 1917.

7. Jennie Franklin and William B. Kennedy split up sometime between mid-1891 and mid-1894, when her billings with Edward W. Coleman began. Her episode in Venezuela is recounted in the *Daily News (Denver),* January 13, 1896. Their last booking found to date is in the *Aberdeen (S.D.) Daily American,* January 9, 1915.

8. "Annie Oakley's Vindication," 709.

9. The date of the destruction of Tussauds's Cody effigy is indicated in a letter reproduced in Sotheby's *S. F. Cody Archive Sale* catalog.

10. "There's No Business Like Show Business" by Irving Berlin
© Copyright 1946 by Irving Berlin

AFTERWORD

1. "Wrights Outdone by Briton," was the front-page headline in the *New York Times,* May 5, 1909, p. 1. Concerning Dunne's severance, the same article quoted the *Times (London):* "It is not known what reasons forced Dunne to adopt this serious step, [i.e., his disassociation with the Balloon Factory] which may be fraught with the gravest consequences for this country." In essence, Dunne was being made a traitorous scapegoat.

2. Dunne's later success was described in the *Guardian,* May 30, 1910, p. 8

3. An extensive article on Dunne and his ideas by H. G. Wells can be found in the *New York Times,* July 10, 1927, p. SM3.

BIBLIOGRAPHY

ARCHIVES

Alaska State Museum photo collection

Annie Oakley Center at the Garst Museum, Greenville, Ohio

Autry National Center in Los Angeles, California

Buffalo Bill Museum Collection and Vincent Mercaldo Collection, Buffalo Bill Historical Center, Cody, Wyoming

Circus poster collection, John and Mable Ringling Museum of Art, Sarasota, Florida

Collection of the Trapshooting Hall of Fame. Vandalia, Ohio

George Grantham Bain Collection, Prints and Photographs Division, Library of Congress

McCaddon Collection of the Barnum and Bailey Circus, Department of Rare Books and Special Collections, Princeton University Library

Montgomery County Archives, Norristown, Pennsylvania

Samuel F. Cody Collection, Drachen Foundation, Seattle, Washington

State Library of Pennsylvania, Harrisburg

W. Duke & Sons Co. Collection, Rare Book, Manuscript, and Special Collections Library, Duke University

NEWSPAPERS AND PERIODICALS

Aberdeen (S.D.) Daily American

American Lawyer

Amusing Journal

Anderson (Ind.) Daily Bulletin

Athens (Ohio) Messenger and Herald

Auburn Citizen

Bangor (Maine) Whig and Daily Courier

Bell's Life

Birmingham (U.K.) Daily Post

Bismarck Tribune

Boston Daily Advertiser

Bristol Mercury

Brooklyn Eagle

Bucks County (Pa.) Gazette
Butte Daily Miner
Camden (N.J.) Post Telegram
Chicago American
Chicago Tribune
Cycling
Daily Inter Ocean
Daily Missoulian
Daily News (Denver)
Dallas Morning News
Decatur (Ill.) Morning Review/Decatur Review
Detroit Free Press
De Olympische Schietsport
Dubuque Telegraph-Herald
Dunkirk (N.Y.) Observer and Evening Observer
Enquirer (Cincinnati)
Field and Stream
Fitchburg (Mass.) Sentinel
Fort Wayne (Ind.) Morning Journal
Galveston Daily News
Geneva (N.Y.) Courier
Graphic (London)
Great Falls Tribune
Guardian
Hamilton (Ohio) Daily Democrat and Evening Democrat
Hartford Courant
Hopkinsville Kentuckian
Indiana State Journal
Iowa State Reporter
Jackson's Oxford Journal
Kansas City Star
Leeds Mercury
Lima (Ohio) Times-Democrat
Liverpool Mercury
Lloyd's Weekly London Newspaper
London Daily News
London Era
Macon (Ga.) Telegraph
Manitoba Daily Free Press
Mansfield (Ohio) News
National Police Gazette
New York Clipper
New York Sun

New York Times
Newark (Ohio) Daily Advocate
Newcastle Chronicle
Newport (R.I.) Mercury
Oakland Evening Tribune
Observer (London)
Omaha World Herald
Oshkosh (Wisc.) Daily Northwesterner
Ottumwa (Iowa) Courier
Paducah Sun
Pall Mall Budget
Pall Mall Gazette
Penny Illustrated
Philadelphia Inquirer
Philadelphia North American
Philadelphia Press
Reading (Pa.) Eagle
Reading (Pa.) Times
Reno Evening Gazette and Reno Weekly Gazette
Reynolds's Newspaper
Rocky Mountain News (Denver)
San Antonio Daily Light
Scotsman (Edinburgh)
St. Louis Daily Globe-Democrat
Syracuse Morning Standard
Times (London)
Van Vert (Ohio) Republican
Washington Post
Washington Times
Waterloo (Iowa) Daily Courier
Waterloo (Iowa) Daily Reporter
Watertown (N.Y.) Daily Times
Waukesha Freeman

SECONDARY SOURCES

"Annie Oakley's Vindication." *Field and Stream* 74(13): 709.

Barton, Francis A. *Jack of All Trades: Reminiscences of an Active and Varied Life.* London: John Bale, 1938.

Broomfield, George Alexander. *Pioneer of the Air: The Life and Times of Colonel S. F. Cody.* Aldershot, U.K.: Gale and Polden, 1953.

Cody, Samuel F. "The Kite That Lifts a Man." *Pearson's Magazine* 16, no. 8 (1903): 106–13.

Coombes, Annie E. *Reinventing Africa: Museums, Material Culture, and Popular Imagination in Late Victorian and Edwardian England.* New Haven: Yale University Press, 1994.

Cortwright, David T. *Dark Paradise: Opiate Addiction in America before 1940.* Cambridge, Mass.: Harvard University Press, 1982.

Dennett, Andrea S. *Weird and Wonderful: The Dime Museum in America.* New York: New York University Press, 1997.

Erskine, Gladys S. *Broncho Charlie: A Saga of the Saddle.* New York: Thomas Y. Crowell, 1934.

Finch, Ralph. "Who's on First? Portlock, Paine, Moreson?" *On Target!: The International Journal for Collectors of Target Balls,* Winter 2009, 13–19.

Georgian, Richard A. "The Luella Forepaugh-Fish Wild West Show 1903." *Bandwagon: The Journal of the Circus Historical Society* 50, no. 4 (2006): 3–8.

Glanfield, John. *Earls Court and Olympia from Buffalo Bill to the "Brits."* Stroud, U.K.: Sutton, 2003.

Hall, Roger A. *Performing the American Frontier, 1870–1906.* New York: Cambridge University Press, 2001.

Hamilton, Richard. "The Sporting Event That Captivated America in the Spring of 1883." www.traphof.org/carver-vs-bogardus.htm (accessed March 5, 2008).

Herlihy, David V. *Bicycle: The History.* New Haven: Yale University Press, 2004.

Hodgson, Barbara. *In the Arms of Morpheus: The Tragic History of Laudanum, Morphine, and Patent Medicines.* Buffalo, N.Y.: Firefly Books, 2001.

Hoig, Stan. *The Oklahoma Land Rush of 1889.* Oklahoma City: Oklahoma Historical Society, 1984.

———. *Cowtown Wichita and the Wild, Wicked West.* Albuquerque: University of New Mexico Press, 2007.

Hornaday, William T. *The Extermination of the American Bison.* Available online at Project Gutenberg www.gutenberg.org/etext/17748 (accessed March 15, 2008).

Hoyt, Harlowe Randall. *Town Hall Tonight.* Englewood Cliffs, N.J.: Prentice-Hall, 1955.

Jenkins, Garry. *Colonel Cody and the Flying Cathedral: The Adventures of the Cowboy Who Conquered the Sky.* New York: Picador, 2000.

John R. Whitley: A Sketch of His Life and Work. London: Dryden Press, 1912.

Julin, Thomas R., and D. Patricia Wallace. "Who's That Crack Shot Trouser Thief?" *Litigation* 28, no. 4 (2002): 1–7.

Kasper, Shirl. *Annie Oakley.* Norman: University of Oklahoma Press, 1992.

Lee, Arthur Stanley Gould. *The Flying Cathedral: The Story of Samuel Franklin Cody, Texan Cowboy, Bronco Buster, Frontiersman, Circus Sharpshooter, Horse Track Racer, Showman, Barnstormer, Man-Carrying Kite Inventor, and Pioneer British Aviator.* London: Methuen, 1965.

Leslie, Thomas. *Iowa State Fair: Country Comes to Town.* New York: Princeton Architectural Press, 2007.

Lowe, Charles. *Four National Exhibitions in London and Their Organizer.* London: T. F. Unwin, 1892.

McMurtry, Larry. *The Colonel and Little Missie: Buffalo Bill, Annie Oakley, and the Beginnings of Superstardom in America.* New York: Simon and Schuster, 2005.

Meyn, Susan L. "Who's Who: The 1896 Sicangu Sioux Visit to the Cincinnati Zoological Gardens." *Museum Anthropology* 16, no. 2 (1992): 21–26.

———."Cincinnati's Wild West: The 1896 Rosebud Sioux Encampment." *American Indian Culture and Research Journal* 26, no. 4 (2002): 1–20.

Ogden, Dunbar H., Douglas McDermott, and Robert K. Sarlós. *Theatre West Image and Impact.* DQR Studies in Literature 7. Amsterdam: Rodopi, 1990.

"Observation War Kites." *Scientific American* 88, no. 24 (June 13, 1903): 445.

Penrose, Harald. *British Aviation: The Pioneer Years, 1903–1914.* London: Putnam, 1967.

Reese, Peter. *The Flying Cowboy Samuel Cody, Britain's First Airman.* Stroud, U.K.: Tempus, 2006.

Riley, Glenda. *The Life and Legacy of Annie Oakley.* The Oklahoma Western Biographies No. 7. Norman: University of Oklahoma Press, 1994.

Roberts, Jean. "Cody—Diary." More Wind for Cody website. www.smaga.de/ecrono.htm (accessed March 14, 2008).

Robertson, John E. L. *Paducah: Frontier to the Atomic Age.* Charleston, S.C.: Arcadia Publishing, 2002.

Schoch, Richard W., ed. *Victorian Theatrical Burlesques.* Burlington, Vt.: Ashgate, 2003.

Shepherd, Dolly, Peter Hearn, and Molly Sedgwick. *When the Chute Went Up—The Adventures of an Edwardian Lady Parachutist.* Oxford: Isis, 1985.

Sotheby's. The S. F. Cody Archive Sale LN6046 Day of Sale, Wednesday, 24th January, 1996, at 10.30 A.M. in the Colonnade Galleries, 34–35 New Bond Street, London WI. Auction catalog. London: Sotheby's, 1995.

Stevenson, E. L. "These Carver Yarns." *Outdoor Life* 55, no. 2 (July 1930): 84.

Stotesbury. Louis. ""The Famous 'Annie Oakley' Libel Suits." *American Lawyer* 13, no. 9 (September 1905): 391.

Thirty-Fourth Annual Report of the Secretary of the State Board of Agriculture of the State of Michigan. Lansing, Mich.: Robert Smith and Co., 1986.

Thorp, Raymond W. *Spirit Gun of the West: The Story of Doc W. F. Carver.* Glendale, Calif.: Arthur H. Clark Co., 1957.

Warren, Louis S. *Buffalo Bill's America: William Cody and the Wild West Show.* New York: Alfred A. Knopf, 2005.

Whitaker, Robert. *Mad in America: Bad Science, Bad Medicine, and the Enduring Mistreatment of the Mentally Ill.* Cambridge, Mass.: Perseus, 2002.

Wiley, Frank W. *Montana and the Sky: The Beginning of Aviation in the Land of the Shining Mountains.* [Helena]: Montana Aeronautics Commission, 1966.

INDEX

Page numbers in italics refer to illustrations.